Idea to Marketplace
An Inventor's Guide

Idea to Marketplace
An Inventor's Guide

by
Thomas R. Lampe

PRICE STERN SLOAN
Los Angeles

Copyright © 1988 by Thomas R. Lampe
Published by Price Stern Sloan, Inc.
360 North La Cienega Boulevard
Los Angeles, California 90048

All rights reserved. No part of this book may be reproduced or transmitted in any form or by any means, electronic or mechanical, including photocopying, recording, or by any information storage and retrieval system without the written permission of the publisher.

Library of Congress Cataloging-in-Publication Data

Lampe, Thomas, R., 1937—
   Idea to marketplace.

   Includes index.
   1. Patents—United States.   2. Intellectual property—United States.   I. Title.
T223.Z1L36   1988   608.773   87-18098
ISBN 0-89586-602-1

Manufactured in the United States of America

10 9 8 7 6 5 4 3 2 1

First Printing

ACKNOWLEDGMENT

I am most grateful to Mike Larsen for his assistance, perseverance, and faith in this project. Thanks also to Sir Jimmy for giving me the time to devote to it.

CONTENTS

Foreword ——————————————————— 9

1. **Follow That Truck** ———————————————— 11
   An Overview

2. **Finding a Need** ————————————————— 27
   The Quest for Ideas

3. **Filling the Need** ————————————————— 39
   Solving the Problem

4. **Don't Just Take My Word for It** ————————— 61
   Proving that the Idea Is Yours

5. **What in the World?** ————————————————— 75
   The Patentability Search

6. **Everything You Wanted to Know about a Patent and Were Afraid to Ask** ———————————————— 91

7. **What's Mine Is Yours—Or Is It?** ————————— 127
   Assignments, Licenses, and Infringements

8. **If You Have an Infinite Number of Monkeys** ——— 141
   The Copyright

9. **Other Bow Strings** ————————————————— 161
   Additional Types of Legal Protection for Ideas

10. **Take My Idea . . . Please** ————————————— 177
    Submitting Your Idea to a Company

11. **Other Options** ——————————————————— 201

12. **The Dream Merchants** ———————————————— 219

    Appendix ——————————————————————— 231

    Index ————————————————————————— 235

FOREWORD

As former chairman of Nikon, Inc., and an author and technical adviser in the field of photography, most of the ideas I've been involved with over the years have, quite naturally, been related to that subject. Photographic ideas, like ideas in all other fields, are products of people's imaginations. Now, not all of the photographic ideas I've seen went on to be great successes. In fact, most didn't. But those that did achieve success came from people who had two things going for them besides a good idea. One is perseverance. The other is knowledge: knowledge not only concerning the idea itself, but about other things such as legal rights and procedures, marketing approaches, the options available to a person with an idea to protect and commercially exploit—the kinds of things *Idea to Marketplace* shares with you.

The history of photography is a history of ideas. Most of us are familiar with the old Chinese proverb "One picture is worth a thousand words." Today one picture is worth many fewer words; in fact, we all know that pictures, symbols, and signs are used to replace even one word. Pictographs such as the skull and crossbones to indicate "Poison," the circle and diagonal bar to indicate "No," and countless others are used in our daily lives. In addition, with the advent of the latest generation of foolproof cameras, the automatic exposure, autofocus, autoflash, autoload, autowind marvels, everyone young and old is a consummate photographer. The pictures from these cameras are sharply in focus, are in full color, and can show an amazing amount of detail. I think our Chinese sages, if they had photos like these, might have said, "One photograph is worth more than a thousand words." Or, with photos of this type an everyday occurrence, they might have skipped recording any type of proverb about pictures and words.

The point is that we tend to accept marvels like photography, color films, and auto-everything anything as ordinary, everyday items that have been around forever. But if we tried to bring some perspective into the science of photography, for example, we would find that photography is about one hundred years old. Color films like Kodachrome are only a little over fifty years old, and these super-automatic cameras are less than ten years old.

Today, development work is being done by many people, companies, and institutes on the next generation of photography. Electronic photography is not a dream any longer. It is here now, not in its final mode, but here. Compare these two scenarios: You're having a family birthday party. Using conventional photographic techniques, you would load your camera with color film, take twenty-four or thirty-six pictures, have the film developed, and then bring the prints home. You would mount them into an album along with hundreds of others, and whenever anyone wanted to see the photos you would bring out the album, with one or two people at a time being able to view them. Now, take the same scenario, but substitute electronic photography. You load your camera with a standard fifty-frame magnetic mini-cassette. You take your pictures and immediately unload and slip the magnetic mini-cassette into a small attachment on your television set to see immediately, in full color, all the pictures just taken. Everyone present can see them at the same time. If someone wants a set of pictures to take home, you quickly make a duplicate set on your television attachment.

When a particular idea or discipline evolves, it doesn't take much further effort to develop and expand the idea in all sorts of directions, much as the example Thomas Lampe gives in Chapter Two for the computer industry. Some of these become valid, workable products or ideas; others are merely pipe dreams or excursions of fancy. You don't know where your imagination will take you, but this imaginative flight must be the first step.

This explosion of ideas and rapid development of equipment and materials that has taken place in photography is just an example of what can and does go on in every single other field imaginable. And this is where the book you are holding comes in. Many readers of this book undoubtedly have ideas, and some of these ideas may potentially be ground-breaking. *Idea to Marketplace* will help you to organize your plans, to protect your ideas, and perhaps to succeed in marketing your product to consumers.

I wish you much luck and success with your ideas, and I hope you have as much fun with them as I did with the photographic innovations I've been involved with over the years. This book will steer you in the right direction.

Joseph C. Abbott
San Rafael, California

1

Follow That Truck
An Overview

FOLLOW THAT TRUCK
AN OVERVIEW

THE BIG QUESTION

I puzzled over the question for years. What makes an idea a financial success? I puzzled over it because my clients puzzled over it. They had good ideas—some had great ideas—but they were interested in the bottom line. Did these ideas have what it took to be money-makers? I was their patent lawyer and my clients had come to me for answers. For me, this was the toughest question of all because it was one I couldn't answer. Somehow my nervous squirming and shuffling of papers didn't quite make it as a satisfactory response.

The question became a bit of an obsession with me. I raised it with colleagues and I found that they too nervously squirmed and shuffled papers. Although happy I wasn't alone in my misery, I continued my search for an answer. I wanted to know what it was that separated the phenomenally successful Pet Rock fad from a beach pebble. Why was one a huge money-maker and the other worthless? I wanted to know.

Then one day, while on a business trip, I ran into an elderly patent attorney of my acquaintance, let's just call him Lawyer Jones. I asked him the question. What makes an idea a financial success? Lo and behold, he had an answer—a unique answer—but before I tell you what it is, let me give you a little background about the old boy.

Lawyer Jones is a man of few words, and he has always chosen those words very carefully. As soon as he gets to the point where a thought can be conveyed by three sentences, he works with renewed diligence to squeeze it into two. Lawyer Jones particularly likes terse answers to questions clients most frequently ask. He'd much rather spend his time at the track participating in the sport of kings.

Lawyer Jones knows that there really is no simple answer to the formidable question of what it takes to make a commercially successful idea. A proper answer might take hours. If he chose to, he could launch into a lecture touching on economics, marketing and manufacturing considerations, the present state of competing technology, and all those other factors, seemingly including the phases of the moon and geopolitical concerns, that go into determining the commercial success or failure of an idea. But he chose not to. The siren calls of the nags at the track wouldn't let him. Therefore, his usual reply was simply, "It depends."

Now, if conciseness were the only standard by which answers were to be judged, this is a truly great one—two words and nine letters. Not bad. Fortunately, however, Lawyer Jones is a very perceptive man and he soon recognized that most of his clients found this answer to be somewhat less than satisfactory. It was a real problem.

Then one day as he was strolling back to the office from a friendly lunch-hour game of single-syllable-word Scrabble with like-minded cronies, the aged patent attorney paused alongside a construction project. The building site was alive with activity. Men and machines moved about in the apparently aimless yet organized fashion that most of us find so fascinating about construction projects and anthills.

Just as he was ready to move on, Lawyer Jones spotted it—a concrete mixer disgorging its contents into a trench. He knew right down to the tips of his arthritic toes that he had found the answer he was looking for. He ran (in a manner of speaking) back to his office, picked up his camera, and returned to the building project. A quick exposure, a speedy trip to the film developer, and a few days later a handsome photograph was framed on his wall.

Now when the tough question comes up, Lawyer Jones merely replies with a smile, a lift of his bushy eyebrows, and a jab of an index finger toward the photograph of the concrete mixer. Prominently displayed on its side is the motto, "Find a Need and Fill It"—a guiding principle attributed to the late Henry Kaiser, Sr., a true innovator if ever there was one.

I, too, now subscribe to that motto. With a bit of help and perseverance, an idea pretty much determines its own value when launched on sometimes cruel commercial seas. And the ideas most likely to succeed are those that best fill needs.

For example, Edward Lowe of Cassapolis, Michigan, recognized a need when he was a young man working for his father hauling sawdust and dried-out clay to gasoline stations. As Mr. Lowe relates the story, a neighbor came by one day and complained about the difficulties he was having with his cat.

Apparently the cat preferred to answer calls of nature in sand, wherever it was to be found. This was not only tough on the neighbor, who had trouble keeping the cat from tracking sand around the living room, but on local kids as well, who were finding unpleasant surprises in their sandboxes.

Clay, Mr. Lowe thought. Clay. The light bulb so beloved by cartoonists practically appeared over his head. It lit up his idea—clay. The gasoline stations used clay to absorb oil spills. Why couldn't it absorb cat refuse? Not only that, but clay is heavier than sand and less likely to get tracked around. The neighbor tried the suggestion, liked it, and spread the word to friends. Mr. Lowe was on his way to becoming the king of cat litter with his "Kitty Litter" brand of products. It goes to show that filling even the most homely and commonplace needs can lead to success.

STRIKING PAY DIRT

This book is a comprehensive and practical guide for people who wish to develop potentially money-making ideas, protect those ideas, and commercially exploit them. I will give you the benefit of things I've learned in almost twenty-five years of practice as an intellectual-property lawyer. Your journey through this perhaps unfamiliar territory will be as painless as possible. It will even be enjoyable. I will use nontechnical terms everyone can understand. This way you won't miss anything along the way.

Getting back to Lawyer Jones, I part company with my old friend, at least a little bit, when it comes to the issue of succinctness. So the first part of this book is devoted to providing you with considerably more guidance on how to generate and develop potentially money-making ideas. A picture is not always worth a thousand words.

I will also show you how to evaluate those ideas on your own, objectively, so you won't waste a lot of time and money pushing a reinvented wheel up a dead-end commercial street. You can then use that creative mind of yours to generate an idea that may be a better candidate for success.

Why do I feel confident enough to make this last statement? Simply because everyone has the inner resources to develop lots of good ideas. The trouble is, most of us don't recognize potentially valuable ideas when they pop into our heads. You will recognize these mental nuggets after reading this book. You will also learn how and where to go prospecting for them.

Okay, carrying this analogy a bit further, what happens after you've struck pay dirt? You have built, or at least thought of, the proverbial better mousetrap. What do you do now? This book answers all of these questions for you. I'll show you precisely how to go about protecting an idea. You will also get advice about how to sell it or how to exploit it commercially in some other way.

There are, of course, certain steps to take to ensure that your idea is legally protected to the fullest possible extent. I cover these, too. This book describes the various types of legal approaches available for protecting ideas. Perhaps your idea is able to be patented or protected by copyright; perhaps it isn't. If it can be patented or copyrighted, you should have a basic knowledge of the patent and copyright laws and the procedures involved, and they are all right here.

If your idea is not patentable or a patent is not desired, or if a copyright isn't applicable to the situation, perhaps there are other forms of protection available to you. These are described. Like people, ideas come in all shapes and sizes, and the law has provided different forms of protection to accommodate this variety. As an idea person you should know about them.

It is most important that you be aware of legal rights and remedies since failure to take certain required steps in a timely manner can seriously jeopardize any inherent property rights you may have to your idea. For example, rights to a patentable invention will be irretrievably lost under certain conditions if a patent application is not filed within a designated time period. A chance to copyright something can also be tossed away forever by not doing the correct thing at the proper time. I'll show you how to avoid that problem.

Let me give you an actual example of how a valuable creative property right can be lost for good. Some years ago, a steamy novel hit the market. It was a work that preachers consigned to hellfire, along with its author. Naturally the book became a bestseller—a hot property, but not the kind the preachers had in mind. Everyone concerned was going to make lots of money, and this was not lost on others. Unauthorized copies started to appear. No problem, thought the author. Sue the knockoff artists for copyright infringement, right? Wrong! As it so happened, there had been inadvertent publication without a copyright notice. This resulted in an invalid copyright under the law existing at the time. That meant that ANYONE had the right to publish the book, and they did. And not one cent of royalties for those copies needed to be paid to the author!

Selling an idea, be it an invention, a copyrighted item, a trade secret, or just a garden-variety idea you can't easily categorize, has its own challenges. How do you find a likely prospective customer for your idea? How do you disclose your idea to another party, such as a company, and at the same time protect your legal interests? Should you use the services of idea and invention promotion or development companies? Where can you go for marketing assistance? Should you commercially exploit the idea on your own? The answers to these and many other questions are found in this book.

Most people dealing with ideas and inventions for the first time harbor a great many misconceptions. There is, for example, the commonly held, false belief that rights to an idea are protected by mailing a letter describing it to oneself. This and other widely held myths and misunderstandings are explored, and exploded. At the risk of repeating myself, there are few areas of the law where rights can be irretrievably lost more quickly because a person fails to take the right steps at the right time.

I hope this book has a side effect. I want it to encourage you to do something with your idea and not let it take up residence in that limbo reserved for things you mean to do but hardly ever get to. Many people often fail to follow up on an idea because they simply don't know what to do with it. Because the law of ideas and inventions is replete with folklore, people are often confused and deterred by misinformation picked up from no doubt well-intentioned friends and associates. Also, I think that there is a natural fear in most of us that keeps us from venturing into unfamiliar areas. And perhaps we fear that our ignorance will be taken advantage of by the many ripoff artists of the world.

I also hope that this book, by dispelling these clouds of misinformation and by arming you with a basic understanding of both the law and practical aspects of ideas and inventions, will encourage you to move forward.

Don't be intimidated by the existence of Big Industry, Big Government, Big Education, and all those other Big Somethings out there with vast research departments and large budgets. The individual with a good idea still has a very important role to play in our country's technological and commercial life. Fortunes continue to be made by people just like you, based on ideas that they have developed. A great many patents are still issued by the U.S. Patent and Trademark Office on inventions made by independent inventors. Thousands upon thousands of valuable copyrights and trademarks are owned by individuals.

And by no means should you be deterred by the fact that you are only one person with an idea. All good ideas evolved from an isolated creative spark generated by a single human brain. Companies such as Apple, Eastman Kodak, Wang, Xerox, and many, many others were built on the foundation of one person's idea.

Take the story of Ruth Handler, for example. She developed the Barbie Doll back in 1959. At that time there were no "young adult" dolls on the market. Ruth felt that there was a need for such a thing after watching her small daughter, Barbara, play out the role of a teen with paper dolls. Why not three-dimensionalize the concept, Mrs. Handler reasoned. Why not, indeed! Just think of the size and scope of the industry hatched from Ruth Handler's simple idea.

But the lady was far from through. Facing retirement from Mattel Corporation years later, Ruth Handler was struck by another crisis—the necessary surgical removal of a breast. Unable to find a suitable artificial replacement, she developed one! Mrs. Handler's "Nearly Me" artificial breast prostheses are now sold nationwide. Mrs. Handler hated retirement and says that her new idea actually saved her life by giving it a new direction. Talk about making lemonade when you're handed a lemon!

Gary Murray is another person who turned tragedy into triumph. The Nevada man is successfully marketing his LUV-BELT, a restraining harness which protects pets while riding in a car—sort of a seat belt for Fido or Pussycat. He came up with the idea after his own dog was killed in an automobile accident.

Like athletes, people with good ideas must follow through to make them pay off. Let me give you an example of follow-through actually involving an athlete with an idea. Fran Tarkenton, the former Vikings quarterback, was flying back to Minneapolis one day when he was struck by a thought; I suppose you could say it literally came to him out of thin air. The idea was simple—selling ad space in airline ticket envelopes. Although Tarkenton himself would be the first to tell you that the idea wasn't particularly brilliant, he did recognize its potential and set out to sell the idea to airlines. Soon Delta, TWA, and Pan Am were on board, along with major advertisers such as Exxon and Prudential-Bache. It was Tarkenton's perseverance that got this idea off the ground and in the air. As he himself put it, "The people who make it in the world aren't the most talented or the smartest. The ones who make it are the dogged ones."

Like Tarkenton or anyone else, idea people often need a little coaching and guidance to perfect their skills. This book will help you do exactly that.

Professional Help

Before we move on, I should say a little about the role of professionals in the idea and invention field. The laws in this area are complex and specialized, and at some point a person with an idea or invention will want to seek professional advice.

There are certain things you can do on your own without the danger of compromising any rights in your idea. Preparing a written disclosure of an idea, as described in Chapter Four, is one of these things. Exactly how much further along the road toward protecting and selling your idea or invention you wish to proceed without professional help is a matter for your own judgment, although I will provide you with useful guidance.

I realize that readers of this book have varying degrees of experience and expertise, so it is difficult to lay down hard and fast rules about when to seek professional assistance. As a general guideline, if you are in doubt about your ability and competence in any of the areas and procedures discussed in this book, you should play it safe and seek professional counsel.

You must bear in mind that this book is by no

means an exhaustive study of the law of ideas and inventions and all its ramifications. There are literally hundreds of technical, legalistic volumes on the subject. It is my intent, however, to present the basics to you so that you can handle the idea in an intelligent manner and take certain steps on your own to perfect and protect your legal rights. Without doubt, professional help is required, in most cases, at some stage to ensure that your rights have been secured and your interests adequately protected.

PATENT ATTORNEYS AND AGENTS

When you have decided that professional help is needed, where do you go to get it? Most inventors employ the services of persons known as patent attorneys or patent agents. These are the only people authorized to represent inventors in patent proceedings before the U.S. Patent and Trademark Office. To be admitted to the Patent Office Register of Patent Attorneys and Agents, a person has to comply with regulations laid down by the U.S. Patent and Trademark Office. Among other things, these regulations require a showing that the person is of good moral character and of good repute and that he or she has the legal, scientific, and technical qualifications necessary to render valuable service to applicants for patents. An examination is given by the Patent Office. Those admitted to the examination must have a college degree in engineering, science, or the equivalent.

The Patent Office registers both attorneys and non-attorneys. The former are known as "patent attorneys"; the latter are called "patent agents." Insofar as the work of preparing an application for a patent and processing it through the Patent Office is concerned, patent agents are usually just as well qualified as patent attorneys, although patent agents cannot conduct patent litigation in the courts nor can they perform various services that the local jurisdiction considers the practice of law. The same restrictions generally apply to patent attorneys who are not admitted to the bar of the particular state in which they practice patent law.

For example, a patent agent or a patent attorney not admitted to practice law other than to represent clients before the Patent Office would not normally be allowed to draw up a license or other contract if, as is likely, the state in which he or she operates considers preparing contracts the practice of law. In the United States the various states determine whether or not one is qualified to represent clients in legal areas outside Patent Office practice. While most patent attorneys are members of their state bar, this is not always the case. For example, patent attorneys may move from a state where they can practice law to another state where they are not admitted to the bar. Patent attorneys admitted in their state of practice often use the title "patent lawyer."

A corollary to the above is that lawyers who have been admitted to the local state bar cannot represent you before the Patent Office if they have not been registered by that agency. They are not patent attorneys or patent lawyers.

As I point out later, there are alternatives to the patent system if you wish to derive benefit from an idea or invention. The locally admitted patent attorney is familiar with these alternatives and can offer legal advice to you concerning them, whereas a patent agent or a patent attorney not admitted locally is somewhat restricted as to the advice he or she can render.

Telephone directories of most large cities have in their classified sections headings for patent attorneys, patent lawyers, and patent agents. And many large cities have patent law associations you can contact for names of people who can provide professional assistance. Also, the U.S. Patent and Trademark Office will provide names of patent attorneys and patent agents in your area upon request. The Office publishes a geographically arranged listing of patent attorneys and agents who have indicated their availability to accept new clients. Simply mail your request for information to Public Service Center, U.S. Patent and Trademark Office, Washington, D.C. 20231, or phone (703) 557-INFO.

CAVEAT EMPTOR

Some individuals and organizations not registered to practice before the Patent Office advertise patent searches, invention development, marketing and patent promotion, and related services. You typically see these ads in handyman, craft, and science magazines. Such individuals and organizations cannot represent inventors before the Patent Office, and they cannot give legal advice. They are not subject to Patent Office discipline and control, and the Patent Office will not assist inventors in dealing with them.

You no doubt have heard the legal Latin expression *caveat emptor*, meaning *let the buyer beware*. This phrase has particular application to individuals and organizations holding themselves out as invention or patent promoters, developers, or brokers. While some of these are reputable and do provide a degree of service to an inventor, there are many sharp operators who do very little of real value for the money you pay them. Do yourself a favor. Read the last chapter of this book before you hand any of your hard-earned money to idea developers or promoters. You can do, on your own, many, if not all, of the things they do. Above all, to be forewarned is to be forearmed.

2
Finding a Need
The Quest for Ideas

FINDING A NEED
THE QUEST FOR IDEAS

INVENTING—A CREATIVE ACT

Inventing is a creative act. As a consequence, it is really impossible for me to tell you exactly how to invent something. I cannot tell a would-be songwriter how to compose a new song or a would-be author how to write the Great American Novel. Similarly, I cannot tell the would-be inventor exactly how to go about generating the inventive spark required to begin the process of inventing. As it turns out, though, most of us do in fact generate inventive sparks on occasion, but we either do not recognize them or are too lazy to do anything about them.

In the next few chapters I'll provide certain guidelines you'll find useful in developing creative ideas and then recognizing them. I'll also give you the simple analytical tools needed to evaluate the merits of your brainchild objectively. You can then decide whether it is worthwhile to pursue it further or advisable to put it in the dustbin of discarded ideas and move on to bigger and better things.

WHY DON'T THEY MAKE . . . ?

As Lawyer Jones advised his clients, the answer to the question "What makes an idea financially successful?" involves a double-barreled answer: Finding a need. Filling that need.

Finding a need in this day and age is actually much easier than you might at first expect, even though modern life already appears full of conveniences seemingly catering to every whim. And yet, more and more new ideas and inventions see the light of day—every day. How does this happen?

Oddly enough, the commercialization of a new idea seems to beget other ideas. Fulfillment of a need actually creates new needs that must be filled. Our technological system functions as an ever larger incubator for new ideas.

Let me give some illustrations. Some years ago a man named Chester Carlson, a patent attorney, recognized a need in the copying industry for a copier that could make clean, fast, dry copies on plain paper. Mr. Carlson's efforts and those of others building on his idea created Xerox Corporation and the xerographic copying process now used throughout the world to create millions of reproductions daily. This development created new needs in the form of improved equipment to handle the process, chemicals and printing materials used in the process, accessories, and so forth. Today millions of dollars are spent annually by corporate and other research departments to satisfy these new needs—needs that did not even exist a couple of decades ago.

And what about the computer? Here again the lesson repeats itself in spades. I lived across the street from an entrepreneur who became a multimillionaire in very short order by developing and marketing anti-glare screens for computers, a relatively simple product, but one for which there was—I can't help myself—a glaring need. This gentleman was in the venetian blind business, making an adequate living, but not exactly qualifying for an appearance on "Lifestyles of the Rich and Famous." Then one day he made a sales call on a large government facility, one having many work stations with personal computers. The operators were having problems with sun glare on the screens. There was a real need for venetian blinds on the windows. That's why he was called to display his wares in the first place.

But my ex-neighbor also recognized another need—a bigger need—one with a solution that was going to be a lot more lucrative than venetian blinds. Why not, he thought, manufacture anti-glare screens for the computer terminals themselves? There were lots of other venetian blinds on the market and there were lots of venetian blind salesmen scrambling to make a living. On the other hand, he didn't know anyone selling anti-glare screens for computers. Our entrepreneur came up with a couple of simple anti-glare screen designs and, voilà!, a successful company was founded.

The principle of the foregoing examples holds true for many other things we use in the course of our everyday lives. The new things that we see about us don't limit inventive opportunities; they *increase inventive opportunities* that we can benefit from.

Many of us have expressed to ourselves the thought beginning with the phrase "Why don't they make . . . ?" when we meet a bit of a problem or need that does not adequately seem to have been solved or met. This thought signals an inventive opportunity, but I'm afraid most of us usually then dismiss the matter from our minds and proceed with whatever we were doing before the thought developed. Either we "let George do it" or, often mistakenly, assume that some company has already solved the problem and we just haven't yet heard of the solution. The point is that all of us at one time or another generate the creative spark required to initiate potentially valuable property rights in the form of patents or salable ideas. Whether we choose to do something about it is another matter.

The story of Howard Hughes, Sr., the inventor of a rotary drill bit which became the foundation of a vast fortune, is instructive. He is a perfect example of someone recognizing a need and filling it—of facing adversity and overcoming it. Hughes was a wildcatter forced to abandon his first oil well because existing drills could not drill through hard rock. Others would have avoided such sites in the future. Not Hughes. He invented a drill to do the job and probably made more money with his invention than he would have in the oil exploration business.

WHERE TO LOOK FOR NEEDS

More than likely, needs to be filled are recognizable during the course of our normal day-to-day activities. For example, a person whose work is taking care of home and children is most likely to recognize needs pertaining to those daily domestic activities. A person spending eight hours a day in an office or factory is obviously more likely to encounter needs to be filled that arise in these surroundings.

The Rubik's Cube is a job-related invention. Erno Rubik is an architect and interior designer who taught in Budapest, Hungary at a design school. The famous cube was a scholarly exercise aimed at teaching students three-dimensional concepts. He realized he was on to something big when all of his associates and friends asked for their own cube. He patented the device and the rest is history. Royalties are still pouring in, and probably will for a long time.

Many people have hobbies or other areas of outside interest that also may be fertile sources of needs to be filled. For example, I have a friend with a passion for model airplanes. He has pursued the hobby for years. After assembling a kit for a remote-controlled helicopter according to furnished plans, he noticed certain operational difficulties in the constructed model when he flew it. He solved these problems by making changes in the construction and he submitted his idea not only to the model manufacturer but to a model airplane magazine, which published an article based on it. My friend was not interested in receiving money for his idea; he was only interested in making a contribution. The point is, though, that he could have tried to sell his idea to the kit manufacturers if he had chosen to.

The most spectacular example of a hobby turning into big bucks is that of the Homebrew Computer Club, a hobbyist organization launched barely a decade ago by a ragtag bunch who shared the dream of making computers easily accessible to anyone. The ideas exchanged at the pioneering club near San Francisco helped create a $40 billion personal computer industry. Steven Jobs and Steve Wozniak, co-founders of Apple Computer, Inc., were two members of the pioneering club. And in this particular case, what started out as a mere hobby rapidly led to multimillion-dollar incomes. Indeed, but for Homebrew, the Apple personal computer might never have been created. Wozniak himself states that a lot of his work was motivated by his desire to have something to show his fellow members.

Of course, not everyone is up to creating computers. You don't have to be a technical whiz to develop a good idea, but you're most likely to make significant contributions in those areas in which you have some degree of knowledge and expertise. If you are a musician, you won't make a meaningful breakthrough in the field of nuclear science. But, of course, nuclear scientists aren't likely to write hit songs!

Fortunately, however, there are many problems to be solved that do not require a formal education or a high-powered technical background. I can give you two examples of very simple ideas launched recently that made their developers millions of dollars. One was the cardboard, accordion-like sunscreen for the inside of an automobile windshield. You know, the one with the sunglass and a hundred other designs (these have appeared in the Sunbelt states where overheating and discoloration of a car's interior can be a real problem). The other was the little yellow sign announcing "Baby on Board." How simple, yes—but, oh, how popular and lucrative they have become. The two brothers who came up with the Auto Shade screen reportedly made $6 million in the first six months of 1986 alone. The fellow with the yellow signs has sold at least 3 million of the items at $2.00 apiece.

THE AGGRAVATION FACTOR

How many times have you purchased a product that in some manner aggravated you? Often such items can be the most simple things. Your aggravation has its source in a specific defect you see in the product. This is a need. If you come up with a solution, it may be worth money to the manufacturer.

When you locate what to you is a problem, don't mentally dismiss it or file it away in your mind as something to get around to one of these days. Instead, at the earliest convenient time, sit down and analyze the defect and think of solutions for the problem. If it is going to be a while before you have time to think about the problem, make a specific written note to yourself. Put the note in a conspicuous location where it will bring itself to your attention. Creative sparks can disappear just as quickly as they arrive and it's best to make note of these things promptly. A prolific inventor I know makes it a practice to carry a small notebook around with him at all times. He jots down problems to be solved as they come to his attention.

BE A TRENDIE

No, a Trendie is not some sort of deviant Trekkie. Keeping a collection of "Star Trek" episodes is not a requirement for being a Trendie. A Trendie is someone who keeps up with trends and developments. If an idea has any real hope of being an economic success, it should relate to a field trending up—not out. Be aware of what's happening—what is getting hot and what is getting cold. Try to position your idea at the hot end of the scale. Following are a couple of examples of current trends. As a Trendie you can find a whole lot more.

Health consciousness is hot today. There are a lot of needs being filled in this field. For example, many items are being introduced to cash in on people's concerns about disease. Just recently I've seen disposable paper shields for placing over telephone mouthpieces. There are also small personal packets of disposable toilet-seat covers that people can carry around with them. These things did not exist even a couple of years ago.

Another offshoot of health consciousness is the trend toward personal exercise equipment to be used in the home. There are many ways to refine and improve this equipment. Just ask Arthur Jones, the entrepreneur who invented the Nautilus concept. He saw how certain changes could be made in existing workout equipment to make it adjustable and give it more flexibility of use. Jones is now happily ensconced on his sprawling Florida estate counting his millions and the alligators he loves to collect.

Another example of a trend is that toward convenience. Actually, this trend has been going on for a long time and shows no signs of slowing down. People want more spare time. They want the little necessary drudgeries of life to be things that they can accomplish as quickly and effortlessly as possible. There are loads of possibilities here for new ideas.

Take the throwaway concept. This is the age of the disposable product. See what you can think of along this line. Actually, one of the inventions I was exposed to early on in my career as a patent lawyer was a convenience throwaway item—the little plastic pop-up thermometer already imbedded in turkeys and other fowl purchased at the supermarket. The little device is perfect from an inventor's point of view because it's a big-volume item with continuous demand, and each is used only once. There may be a small royalty per thermometer, but, my, how these royalties add up.

I mentioned being aware of developments, as well as trends. Often a technological advance makes an idea feasible when it wasn't before. One of the hottest new toys around could not have existed until very recently because the state of technology didn't permit it to, and yet it's nothing more than a variation of an old game that's been around for years.

The toy is Lazer Tag. I'm told that Lazer Tag came about when one of its inventors was musing about playing cops and robbers as a kid. "Bang, bang, I got you" was the standard language used. What if, the inventor thought, you really did get that other kid. Well, now you can—with a low-energy laser weapon and transponder worn by your opponent, which lets you know when he's been "tagged." Lazer Tag—a great idea that has and will likely continue to put a lot of money in the bank for its developers.

About Fads

I should probably say a few words about fads. You might think that fads don't fill a need; that, in a way, they're exceptions to my little theory that the best ideas fill needs.

Actually, fads do fill a need—the need within all of us to be entertained and amused. Some of the most lucrative ideas have actually been on the faddish side—pyrotechnic rockets that quickly rise to a zenith, burst full bloom to briefly illuminate the commercial skies, and then fall to earth as burned-out cinders, never to be seen again. Take the Pet Rock, for example.

Well over a million people did—from Gary Dahl, its inventor, and he made at least a dollar on every one.

Or think of the Wacky Wallwalker, the eight-legged, sticky rubber toy that made Ken Hakuta a millionaire. He sold close to 150 million of the items before the fad cooled off.

A U.S. Supreme Court justice once said that he couldn't define pornography, but he knew it when he saw it. Fads are like that. They're pretty hard to get a handle on. Hakuta, an expert in the field, certainly knows this. He keeps trying to spot the next big one. It's not easy and, so far at least, no new Wacky Wallwalkers have crossed his desk, despite lots of time and effort on the young entrepreneur's part. He's established a toll-free hotline (1-800-USA-FADS) to receive calls from would-be fad makers, hoping that lightning will strike twice.

Employer Rights

A note of caution. If your idea pertains to your job, your employer might have certain rights to it, either by virtue of an employment agreement many employees are asked to sign upon hiring or, even in the absence of an express written agreement, by virtue of the fact that you thought of the idea or invention on the job. Use of an employer's materials and equipment to develop an idea or invention can also give him rights. Also, if you have submitted your idea or invention in the company suggestion plan you may have given your employer certain rights.

The employer's rights can go as far as outright ownership to "shop rights" that only permit him to use the invention in his business. If you have signed an employment agreement or have other reason to suspect that your employer may have acquired rights in your development, you should engage the services of a patent attorney to find out what your rights are and how best to proceed.

Actually, in many cases, an employer can be very helpful to an employee wishing to exploit an invention. So you really shouldn't look at such a situation as necessarily bad. Perhaps you and the employer can both commercially exploit the idea or invention to your mutual benefit.

So there you have it. Finding a need for a new idea or invention is probably easier than you realized. Just be alert for the presence of the creative thought signaling itself with the words, "Why don't they make . . . ?"

The next step—filling the need.

3

FILLING THE NEED
SOLVING THE PROBLEM

FILLING THE NEED
SOLVING THE PROBLEM

THINGS TO DO

Let's say you've recognized a problem or a need to be filled. You've also thought of a solution. Are you on Easy Street yet? Not by a long shot. There is still a lot to be done. Thomas Edison is reputed to have said that invention is 2 percent inspiration and 98 percent perspiration. This is where the perspiration part comes in. It is also where many would-be inventors drop by the wayside. Don't be one of them. If you have faith in your idea, don't let a little hard work stand between you and financial success.

The first thing you have to do is put your solution down in writing, sign the writing, date it, and have it witnessed. (See the next chapter for an in-depth treatment of how to make a written record of your idea so that you can prove it is yours.) It is now also time to ask yourself some basic questions. Such as: Is the idea new? Will the idea work? Is it commercially and financially feasible?

IS YOUR IDEA NEW?

To me, the most fundamental question you should ask yourself is whether your idea is in fact new. Why waste your money and effort on something which is little more than a clone of something already in existence? Believe me, this happens. I had a new client appear in my office one day with an elaborate (and obviously very expensive) working model of a furnace he had devised. This labor of love was hooked up to a number of instruments which showed just how efficient the furnace was. And it WAS efficient. The only problem was that it wasn't new. There were furnaces on the market operating just like his. He just hadn't seen them. He hadn't bothered to look—afraid maybe. His invention had taken on a life of its own and he just kept plowing ahead on something so clever he believed that only he could have thought of it. As the saying goes, it's terrible to see a grown man cry.

We found out that the furnace design was old by conducting a search of U.S. Patent and Trademark Office records, and this is certainly one excellent way to find out if something has been invented before. After all, the Patent Office is the greatest repository of technical information in the world. Chapter 5 of this book tells you all about patent searches, even how you can conduct one on your own.

There are other ways to find out whether something is new, and you should use them. After all, not everything in the world shows up in patents. Some things can't be patented. Other things may have been patentable but the inventors never got around to doing it. And, of course, a patentability search is not an exact science. Pertinent patents can be (and are) missed in searches, no matter how expert the job.

Here are some of the investigations that you can do yourself.

Check your local stores, especially those you feel would likely be an eventual sales outlet for your proposed product. For example, if you've developed a new flashlight especially designed to be used by campers, check out not only camping supply stores, but also hardware stores and stores like Sears carrying a full range of merchandise. There are a lot of people and companies constantly thinking and innovating and it is not unusual for a would-be inventor to find that the very thing he or she has in mind, or something very close to it, is already on the market. It may even be that such a product has been a long-term fixture on a sales counter passed many times on the way to other locations in a store visited often. I am afraid many of us look, but we do not see when we have something else on our minds.

If you do not locate the type of product you have in mind, check with the sales personnel in the store. You don't have to tell them you've invented a specific product and would like to see if they have something like it. Some innovators are reluctant to ask this type of question since they don't want to tip their inventive hand. Also, as you will see later, disclosure to someone of details of an idea can be harmful to legal rights, especially those under the patent law. You can ask a question in more general terms and still obtain the kind of information you wish. For example, if you have a concept for a new household appliance, say, a vegetable peeler with an anti-clogging feature, it is not necessary (or desirable) to divulge its specific construction when asking your question. You might simply ask the salesperson if there is something on the market performing the same general function of peeling a vegetable without clogging.

There are other sources from which you can

obtain information about potential competing products. For example, many manufacturing companies have sales and trade catalogs that they will send on request. Call or write these companies. Get their names and addresses from the *Thomas Register* at your library. This register lists companies by product category. The large merchandising companies having extensive mail-order operations, such as Sears and Montgomery Ward, of course have catalogs describing many thousands of items. This is an excellent way to spot competing products sold to the general public. If your proposed new product is in a specialized area serviced by specialty magazines, such as hobby magazines, ads found in these publications are often valuable sources of information.

Even a manufacturer that doesn't have a catalog can be helpful in your investigation. You can write, describe your product in very general terms, preferably merely by its function, and ask whether the manufacturer makes available anything like it. In your letter you might ask the company to suggest others to contact if it does not carry such a product.

Please be very careful here though. There is always the danger that a manufacturer might not be making such a product, but that your contact will prompt it to do so. You could be creating a competitor with your letter. Also, disclosure of an invention before filing a patent application has potentially harmful legal consequences. I'll get into these later in this book. Some inventors follow the practice of not making any public disclosure of their invention until a patent application is actually filed.

Will Your Idea Work? What About a Model?

I encourage construction of a model by my own clients if there is any question whatsoever as to an invention's workability. After all, even inventive geniuses can have problems converting a great idea into something that actually works. James Watt, for example, designed the world's first steam engine in only two days. It took him ten years to produce the first working model.

In addition to proving out operation, models have many other advantages. They often make good sales tools. In addition, there are often hidden problems that will come to light only after the model has been constructed and operated. It is very important that you recognize such problems early in the game. Then you'll have time to solve them before filing a patent application on the invention or trying to sell it.

Making a Model

Models of some inventions and ideas can be built by you, the inventor, especially if you are handy with your hands and have access to the necessary tools. On the other hand, some things are of such complexity or involve such specialized machining or other work that it is wise to contact people in the model-making business. The Yellow Pages of most big-city telephone directories lists such people under headings like "Model Making," "Pattern Makers," or "Industrial Designers."

If you are interested in having your model made by someone in this line of business, pay a visit with your plans or drawings so that you can discuss the matter. Make your disclosure only after being assured that your idea will be kept confidential. Get this promise in writing. By all means, before you commit yourself to having a model made, get a firm bid, preferably in writing, from the model maker on the price he will charge you. Also, you might ask him to show you examples of his work so you can assure yourself that a good, workmanlike job will be done. Some indication of a model maker's reliability may be gleaned from the number of years he has been in business, as well as from the names of his clients. Chances are, if he has a number of well-established clients, his work will be satisfactory. Don't be shy about asking the model maker these questions. It's your money that will be spent and you are entitled to a good job for it.

The amount of information you will have to provide the model maker about your invention to have an acceptable model made depends on a couple of factors. First, if the invention is highly complex or involves close tolerances, the information will have to be sufficiently specific to provide these things. Second, there is always the question of how much you can rely on the experience and initiative of the model maker. The more initiative he displays on his own and the more experience he has, the less detailed need be the information you submit. You should be able to get an adequate fix on the abilities of the model maker after a conversation or two about your idea.

Although they do not officially designate themselves as model makers or pattern makers, in most communities in the United States, large and small, there are other people equipped to build models. They may be listed in the telephone directory under a wide variety of titles, such as metalworkers, blacksmiths, welders, etc. Finding a person with a title is not nearly as important as finding a person who will be able to do a good job for you. It is just a question of finding them, either by letting your "fingers do the walking" or by making personal visits to prospective model builders. Your patent attorney may be able to provide you with names of model builders.

A couple of words of caution about building a model. I strongly recommend that you *not* go to an invention promotion or brokerage company for advice about making a model, at least until satisfied that you would be dealing with a reputable organization. How do you find out whether an organization is reputable? Well, one place to start is the Better Business Bureau of the city where it's located, or the Bureau of Commerce and Industry or Bureau of Consumer Affairs of a state where it maintains a place of business. You can also ask your patent attorney.

Also, before using the facilities of your employer to build a model or do any other work relative to your idea, be aware of the possible legal ramifications. As mentioned in the preceding chapter, using an employer's tools, materials, etc., could give the employer certain rights to your invention—rights you may not wish it to have.

Developmental Assistance

People sometimes run into snags developing their ideas. It isn't uncommon for someone to come up with a basic idea and then find that the contributions of others with special expertise are necessary to make it really work. For example, an inventor might come up with a product that requires specific electronic circuitry. The inventor knows, or at least believes, that the circuitry is feasible, but he or she simply does not know enough about electronics to devise a specific circuit to do the job. Obviously, to make the invention work, the services of someone with that knowledge will have to be used to complete the job.

There is no reason whatsoever why you should be afraid to seek expert assistance as to certain features of your invention. The chances of someone actually stealing your idea are quite minimal if you have kept and maintained proper records as suggested in the next chapter. You should be aware, however, that if the other party's contribution is of such significance as to make an inventive contribution, that person may become a co-inventor with you of your invention and thus become part owner.

I was involved with a situation like this not long ago. An inventor hired an assistant to help him with a project aimed at developing a pneumatic conveyor system. The intent was that the assistant merely follow instructions—help the inventor assemble and set up the equipment, that sort of thing. As it turned out, the assistant did much more. He made specific suggestions for modifying the equipment and these changes were what made the invention operate properly. The assistant became co-inventor and the application had to be filed in both men's names. The inventor was required to pay the assistant an extra $10,000 to obtain full control over the patented invention.

Whether someone makes an inventive contribution or merely employs ordinary skill in tying up loose ends of your invention is often a close call and I certainly recommend that you contact a patent attorney for advice before retaining the services of someone to complete an invention.

Some people have turned to employers for advice and help when developing an invention, often with the thought in mind that the employer's facilities could be used or that the employer or another employee might make a technical contribution. Again, there is certainly nothing wrong with this as long as you recognize the possible legal consequences.

WHERE TO GO FOR HELP

Private and public research organizations exist to help people develop their inventions. Sometimes an inventor has neither the time nor the technical expertise to complete an invention to the commercial stage. Probably the best-known invention development research organization of this type is the Battelle Development Corporation at 505 King Avenue, Columbus, OH 43201. Batelle will consider inventions submitted to it, and possibly assist in their development and exploitation under an appropriate agreement with the outside submitter. Batelle worked with Haloid Corporation (now Xerox Corporation), for example, to bring Chester Carlson's ideas to commercial fruition.

Other private-sector organizations offering various forms of assistance to inventors are in existence. These include:

>Arthur D. Little Enterprises
>Arthur D. Little, Inc.
>Acorn Park
>Cambridge, MA 02140

Cambridge Research & Development Group
21 Bridge Square
Westport, CT 06880

Gulf & Western Invention Development Corp.
1 Gulf & Western Plaza
New York, NY 10023

HMS Associates Company
2425 Maryland Road
Willow Grove, PA 19090

Model Builders, Inc.
6155 So. Oak Park Avenue
Chicago, IL 60638

National Patent Development Corp.
375 Park Avenue
New York, NY 10022

Product Resources International
90 Park Avenue
New York, NY 10016

Research Corporation
405 Lexington Avenue
New York, NY 10017

Your local Yellow Pages probably lists Product Developers in your area, but it often isn't necessary to go to these expensive specialists for help. Sometimes a little networking with friends and associates will locate someone to help you solve a particular problem on a more informal basis. The model makers I mentioned earlier often have more than enough technical expertise to get the job done for you.

There are a number of government and university organizations which provide assistance to inventors.

Perhaps your state has one. For starters, contact your state universities to find out whether they have such a program. Also, check your telephone directory under "State Government" for listings of agencies with titles such as "Industrial Development Department" or "Department of Labor and Industry," the most likely types of agencies to steer you to inventor assistance programs if they exist. Some governmental and academic agencies available to provide some degree of assistance to inventors are:

>Advanced Technology Development Center
>Georgia Institute of Technology
>430 10th Street, N.W.
>Atlanta, GA 30332

>Alabama High Technology Assistance Center
>Morton Hall
>University of Alabama
>Huntsville, AL 35899

>Baylor University Innovation Evaluation Program
>Center for Entrepreneurship
>Hankamer School of Business
>Baylor University
>Waco, TX 78798

>California Polytechnic State University
>San Luis Obispo, CA 93407

>Carnegie-Mellon University
>Center for Entrepreneurial Development
>4516 Henry Street
>Pittsburgh, PA 15213

>Center for Innovation and Business Development
>Box 8103
>University Station
>Grand Forks, ND 58202

Institute for Ventures in New Technology
Suite 310, Energy Research Center
Texas A & M University
College Station, TX 77843

Invention Assessment Center
180 Nickerson, Suite 310
Washington State University
Seattle, WA 98109

New England Industrial Resource Development Program
Dearborn, NH 03824

New Mexico Energy Research and Development Institute
University of New Mexico
Room 358, Pinon Building
1220 South St. Francis Drive
Santa Fe, NM 87501

Science and Technology Resource Center
Southwest State University
Marshall, MN 56258

Stanford University
Innovation Center
Stanford, CA 94305

Technology Idea Evaluation Program
Missouri Division of Community Economic Development
P.O. Box 118
Jefferson City, MI 65102

University of Kansas
Center for Research, Inc.
2291 Irving Hill Road, Campus West
Lawrence, KS 66045

University of New Mexico
Technical Applications Center
Albuquerque, NM 87131

University of Utah
Utah Innovation Center
Salt Lake City, UT 84112

Wisconsin Innovation Service Center
402 McCutchan
University of Wisconsin
Whitewater, WI 53190

Inventors clubs can be great sources of information about where to go for help. I've listed some of the major organizations of this type in Chapter 11.

IS YOUR IDEA COMMERCIALLY FEASIBLE? HOW PRACTICAL IS IT?

Surprisingly enough, many inventors and idea people do not ask themselves whether their brainchildren are practical. Or, if they do ask themselves that question, they don't get around to it until way too much time, energy, and money have been wasted in development. Actually, this is a question that should be asked in the very early stages of a new product's life.

People develop innovative ideas primarily as a result of two motivations. One, of course, is the personal satisfaction that comes from solving a problem or satisfying a need, as well as the recognition going along with it. I'll bet you can guess the second motivating factor. Money. And to make money, a new development must be commercially feasible. It must be able to compete successfully in the marketplace with a multitude of other products. During all stages of development of a new idea or invention, you must constantly keep this fact in mind. You must continually look at the product itself from the standpoint of its commercial practicality. You must also have an understanding of marketing realities, that is, whether your new development can hope to make it successfully in the real commercial world.

These are not easy questions to answer and even large experienced companies find that new products fail to survive introduction. Nonetheless, you must do your best to try to assess your new development from a commercial point of view. If you don't, a lot of good money can be thrown after bad before the cruel truth hits home.

Just because a product is new doesn't mean it will be commercially successful, nor does the fact that a development is patentable ensure financial returns to the inventor. Patents, copyrights, and trademarks are not necessarily passports to success, although they certainly can be valuable tools for commercializing a worthwhile development under the right circumstances. I'll go into that in much greater detail later on. For the time being, let's forget about the tools for protecting ideas and look strictly at the commercial realities relating to the new development itself, whether protectable or not.

Look at the Product Itself

Companies take a cold, hard look at a new product before deciding whether to manufacture and market it. In industry, this is known as making a "product [or engineering] evaluation." Sometimes these evaluations can be quite lengthy. They may involve elaborate testing procedures as well as detailed investigations of the economics of manufacture.

As a private inventor, you probably don't have

access to a large engineering staff or to testing and other equipment that might be used in making an in-depth evaluation. However, in most cases, these things are not really needed and you can probably make a decent appraisal of your own development without going through an elaborate formal evaluation program. Further, a worthwhile evaluation normally can be made without any cost whatsoever if you use a little common sense and ask yourself a few basic questions about the development.

The first question you should ask and answer I've already covered. "Will it work?" A rather obvious question, but, surprisingly, one many new inventors fail to ask of themselves, only to find out later that a new product has some inherent defect or that assumptions made on paper did not work out in practice.

You must also look at the proposed new product from the standpoint of cost. You must ask yourself whether the product can be produced at a price attractive to prospective purchasers, yet enable the manufacturer to make a profit. You must not only look at the costs of the materials involved in the construction of the unit, but also think in terms of its cost of assembly.

You should constantly attempt to simplify the product; work toward eliminating all parts not necessary to its proper function. This can result in cost savings, and it generally tends to make the revised product more reliable in operation. I think we've all seen examples of things employing space age technology and awe-inspiring complexity that would operate much more reliably and for a longer period of time if only they were simpler. And they don't all come out of the Pentagon either. Some people have a tendency to fall in love with technology. They will use it for its own sake merely because it exists. Never mind that something can be done more simply and cheaper. Well, that kind of thinking is all very well when you're sitting in an ivory tower someplace working on a government grant, but it is not the kind of attitude likely to lead to a marketable product. Any person with a proposed new product might consider hanging a sign in the workshop with the word "SIMPLIFY" spelled out in big block letters.

Finally, insofar as the product itself is concerned, whenever possible, try to incorporate off-the-shelf components rather than something that must be custom fabricated. While this is not always possible, the use of commercially available parts in a product will usually greatly cut down on manufacturing costs and may contribute to its reliability as well.

I had a client once whose invention incorporated a number of relay switches. This man was a perfectionist who didn't want to use commercial switches because he thought they were too noisy. He had his switches custom-made, which added 20 percent to the cost of his product. You have to understand that somebody buying this product didn't give a darn whether the switches were noisy or not. The products were to be used in environments likely to be noisy themselves. And anyway, a nice sturdy housing surrounded the switches. The inventor's decision was based on a personal quirk and that quirk caused his product to be overpriced and thus a commercial loser.

LOOK AT THE MARKET

Worthwhile ideas and inventions cannot be made

in a vacuum. While the classic stereotype of the independent inventor is someone leading a reclusive life in a remote garret, it is highly unlikely that such a person would come up with a new, commercially worthwhile development. You have to know what the market is in order to assess your new idea realistically. Again, a formal study is not necessary to determine how your development might stack up in the marketplace. It is again a matter of asking yourself a few basic questions and attempting to answer them as objectively and accurately as possible.

You will want to know how your proposed product compares with existing competing products. As a practical matter, very few new product innovations are so completely revolutionary that there will not be other products already in the marketplace competing at least to some extent. In fact, an absence of anything remotely similar might be sending a message that there really isn't a need to fill.

Assuming that you find potentially competing products, you will want to decide exactly how yours relates in terms of competition. Compare your proposed new product with existing products from the standpoint of cost, appearance, and reliability. Determine what advantages, if any, your product has over existing products that would enable it to successfully compete against them. Admittedly, a lot of educated guesswork will be involved here since the vagaries of the marketplace are almost beyond human comprehension at times, but it is still a worthwhile exercise.

Cost is not the sole criterion for determining competitiveness of a product. If your product has significant advantages as compared to those already in existence, you may be able to justify a premium price for it.

I recommend that you do not rely entirely on your own reaction for determining your product's potential competitiveness since, as the developer of the invention, you may not be 100 percent objective. Comments of friends and associates whom you trust should be sought to assist you in making an accurate judgment. Find out if they like your idea. If so, why? If not, why? What specifically don't they like about it? How much do they think it's worth? Would they buy it?

THE CONSUMER SURVEY

You can, if you wish, conduct an actual consumer survey, in a more formal way. While such surveys can be fallible (after all, there was the Edsel and, more recently, the attempt to dump the original-formula Coke), the results obtained may often be helpful, not only in assessing the merits of your idea, but as a possible sales tool when you are attempting to sell your brainchild. Surveys can be strong evidence that you have a truly marketable product.

Ross Williams carried out a particularly effective consumer test. Williams and his wife developed a type of lotion-impregnated towel that could be used without water, something they reasoned would be particularly useful in an environment like an airplane. The new product was compact, easy to use, and disposable. Williams bought a ticket on an airline and passed out homemade samples to fellow passengers. Talk about captive audiences! The result was everything the two inventors hoped for. The airline users loved the product, and so did the airline itself, to the tune of an order for a half million towels. This was how the famous Wash 'n' Dri product was born.

A typical survey is conducted by identifying in your own mind the particular group or class of people who would use your idea and deciding upon a representative sample of that class who will take part in the actual survey. For example, if janitors would be the primary customers, you would contact as many janitors as you see fit to give a good response to the sample. This could be any number that you decide upon under the circumstances, although I suggest surveying at least ten people to give a halfway decent cross section.

After showing the idea to the survey subjects you will want to discuss it with them. Find out what the pros and cons are. How does the idea compare to its closest competitor? You might want to pass out a written questionnaire. In any event, you will want to hit the following points about the product:

1. What do you think of its appearance?
2. How much would you be willing to pay for it?
3. How well will it do the job for which it is intended?
4. Do you think it will hold up well under actual use?
5. What about ease of use—any problems?
6. Are you aware of anything else like it? What are the differences?

If possible, allow the survey participants to use the product for awhile so that you can get input of an especially practical nature. Merely looking at something and using it are two different things. Something can look great but work terribly, and vice versa.

Since public disclosure and use of an invention more than a year before filing a patent application will invalidate a patent, you can't run a public survey or test your idea for an unlimited period of time if you anticipate filing for patent protection. If you are interested in obtaining foreign patents, you should definitely hold off altogether until a patent application is filed. In any event, I recommend seeking the advice of a patent attorney before you make any disclosure of your invention to someone else. Your patent attorney will provide you with guidance and assistance with regard to a survey. If you wish to conduct a survey before filing a patent application, you will be told things you can do to protect yourself. For example, your attorney may suggest use of a secrecy agreement to be signed by participants in the survey.

As a final comment, don't be discouraged if your closer look at the product and market situation seems to have dimmed chances of success for your development. You should use the information that you have obtained in a constructive manner. Hopefully, you have received valuable insights that will be helpful to you. If you still feel the idea has potential, use these insights to rework your idea and make it more commercially sound. If, on the other hand, your investigations have led you to reluctantly conclude that your idea is not really new or not commercially worthwhile for other reasons, you can at least stop wasting your time and turn your efforts to something new, exciting, and different. There are many more needs out there waiting to be filled.

4

Don't Just Take My Word for It

Proving that the Idea Is Yours

DON'T JUST TAKE MY WORD FOR IT
PROVING THAT THE IDEA IS YOURS

MAKING A WRITTEN RECORD

It may become important for you to be able to prove the date when you first thought of your idea. This is true in the case of both patentable and nonpatentable inventions and ideas.

When you file a patent application on your invention, you will have to make an oath or legal declaration that you believe yourself to be the original and first inventor, and it may be necessary for you to prove this later on. The date of conception of an invention (that is, the date it is thought of) is one of the factors entering into the determination of who, in fact, is "first inventor" under patent law. There are other forms of protection for ideas and inventions outside the confines of the patent laws. You'll learn about them a little later on in this book. But there, too, it may be necessary to prove that you are the idea's originator, or at least have an earlier or superior claim to it as compared to other people.

As promptly as possible after you have thought of your idea, you should put it down in writing. Be sure to sign and date the description. While there is no particular legal form necessary for recording your idea or invention, it is important to keep in mind that the document at some time or other may have to be introduced into evidence to prove and enforce your rights. There are a few basic rules you should follow to add to the credibility and usefulness of the written record as evidence that you are the originator of the idea described in it.

First and foremost, your description should be clear and concise. Do not take things for granted or leave room for guesswork insofar as your description is concerned. Keep in mind that other people who may read your description later on will not have the benefit of the prior work or research that you may have put into the invention or idea.

Although not absolutely necessary, it is often helpful to start out your description with a brief statement of the problem you are attempting to solve. Let me give you an example. Suppose you have invented a device which lets you know when the mail has been delivered. Your problem statement could read something like this: "People sometimes go to their mailboxes only to find that the mail has not yet been delivered. This is not only aggravating but a waste of time and energy as well. My device solves this problem by automatically sensing when the mail is delivered and signaling when this has occurred."

You will then want to follow up the problem statement with as complete a description as possible, in your own words, of the idea. Getting back to the mail delivery problem, you would want to describe all of the principal parts of the device and how they work. Your device, for example, might include a pressure switch on the bottom of the mail box, the switch being connected by electrical wiring to a power source, such as a battery, and a signaling component inside the house, an electric light perhaps. When the weight of the mail closes the switch, the circuit closes and the light is lit. You consider these parts to be the necessary components of your invention, so you would want to make absolutely sure that your write-up includes them all, along with a description of how they cooperate. If you have several alternative arrangements in mind, be sure you describe them. For example, your write-up might state that the signal could be a buzzer rather than a light, or that a radio transmitter and receiver could be used instead of electrical wiring to send a signal from the mailbox to the house.

If they will contribute to the understanding of your proposal, you should add sketches or drawings illustrating it. These need not be fancy, merely understandable. For example, the sketch for the mail delivery device could be simplicity itself—just a single figure showing a mailbox, the switch on the bottom, and the signal light and batteries electrically connected to it. If you have made a model, take photographs and refer to them in your written description. If drawings are used in your write-up, they should show all essential features necessary for a proper understanding of your idea or invention. If size, shape, and particular types of parts or materials are important, they should be clearly indicated on the drawing, as well as in the written description. If one view is not adequate, employ several views, clearly identifying and cross-referencing them in your written description.

It is not necessary that the description be typed; however, if it is written in your own handwriting, use indelible ink since descriptions made in pencil or other easily erasable or removable writing substances can bring the description under attack on the grounds that it has been altered in some way. If you make a mistake and an entry requires correction, do not erase it. Instead, draw a line though the portion to be corrected and rewrite it immediately below, initialing and dating the correction.

You want to avoid any question being raised that your written description has been altered or supplemented later. This could bring your inventorship into question, if it is ever challenged, and it probably would be if you ever had to sue to enforce your rights. For this reason, it is best to use a bound notebook (not ringbound) with consecutively numbered pages to describe your idea or invention. This is the practice followed by research engineers and scientists of most large companies experienced in these matters. Keep the notebook intact. Do not rip out any of the pages. If the notebook is not kept intact, there will always be some question as to the nature of the material deleted from the notebook. This can destroy or weaken the credibility of the notebook as supporting your origination of the idea. If you do not have a bound record book, the next best thing is to secure together separate sheets and have them numbered consecutively.

WITNESSING THE DESCRIPTION

It is very important to have your description witnessed as promptly as possible after you have written it. You should exercise great care in selecting your witnesses for this purpose. First of all, the witnesses should be trustworthy friends or acquaintances whom you have confidence in. Because it may be necessary for the witnesses to testify in your behalf at some future date quite far removed from the date they witnessed your description, it is advisable to have at least two persons act as witnesses, thereby avoiding sole reliance on one witness who may die, move away, or in some other manner become incapacitated as a witness. To give you an extreme example and to show you how long these situations can be, after a 26-year legal battle, a federal judge ruled that Gordon Gould of Kinsale, Virginia, invented the laser used in such things as compact disc players and bar code readers. Mr. Gould probably considers it worth the wait since he could get $10 million or more a year in royalties from his invention.

A person selected as a witness should be a financially disinterested party insofar as the idea is concerned. Witnesses should not be business associates having a possible financial interest in the idea or invention, nor should they be close members of the family, such as your spouse or children, who stand to be indirect beneficiaries.

A witness who does not understand your idea is really no witness at all. Although the witness need not understand all the ins and outs of the scientific principles, if any, employed in your development, he should know how it is constructed and how it operates. Obviously then, you will have to use some judgment in selecting a witness based on the complexity of your invention. For example, a witness eminently suitable in the case of a relatively simple household gadget may not be suitable at all as a witness when the invention is in a relatively complex field, such as computer technology.

After you have selected your two witnesses, you should have them read your written description to make certain that they understand it. If the witnesses need to ask questions to understand the description, this may mean that you will want to supplement your write-up in order to provide additional clarification of the points raised.

Remember that you want your description to be as complete as possible. It should stand on its own insofar as describing your idea or invention is concerned. Any gaps in your description could be harmful to you at a later date. Often, for example, an inventor will find that he has been using terms or abbreviations that may be clear to him but not to other people reading the description. The comments of prospective witnesses can be very helpful to you in pinpointing such difficulties.

After the witnesses have read the description and have indicated to you that it is clear to them, they should indicate such fact at the end of the description, sign and date it. This is in addition to your own signature and date. I suggest that EACH of the witnesses do this by writing the following notation immediately at the end of the descriptive write-up:

Read in confidence and understood by me this _____ day of _____ , 19____.

(Signature of Witness)

If the invention is complex and its development has taken place in steps over a long period of time, it may be desirable to have the portions of the disclosure document relating to these various steps separately signed, dated, and witnessed over the period of time of the development.

After you have written your description, dated and signed it, and had it witnessed, you should keep this document in a secure place such as your home or office safe or a safety deposit box for possible use by you in proving your rights to the idea or invention at some future date. Many an inventor has lived to regret his failure to do this. I am reminded of the (possibly apocryphal) story of how the corporate employer of Dr. Charles Steinmetz, a prolific designer of electrical equipment, hired a full-time employee to follow the absentminded inventor around his laboratory to retrieve and file all discarded scraps of paper upon which the great scientist had made notes. Unfortunately, most idea people do not have this luxury and must rely on their own efforts to keep their records intact, safe, and sound.

OTHER FORMS OF CORROBORATION

Materials other than a witnessed description showing that an idea was conceived as of a certain date should also be kept in a safe place. These would include, of course, the drawings or photographs referred to in the description and other ancillary items such as a model, purchase orders for parts used in the model, correspondence relating to the invention, preliminary sketches and notes (dated and signed), and any other materials and documents tending to corroborate origination of the idea. I recommend that all of these items be assembled in one place with the signed, dated, and witnessed description.

It may be helpful to include with these materials a list showing all of the things collected together. You might also wish to make and save notes of other facts tending to show that the idea or invention is yours. For example, if disclosures of the idea have been made to people other than the witnesses who have signed the descriptive document, you may wish to prepare a note identifying them and the circumstances and dates of the disclosure.

If you intend to send all or a portion of these materials to others, such as patent attorneys, patent search companies, etc., copies—not the original documents—should be used. It is very easy for things to get lost, misplaced, or simply forgotten once out of your possession.

THE U.S. PATENT AND TRADEMARK OFFICE'S "DISCLOSURE DOCUMENT PROGRAM"

The U.S. Patent and Trademark Office has established a service many people have found useful when proving an idea is theirs. Under this service, called the "Disclosure Document Program," the U.S. Patent and Trademark Office will accept and preserve for a limited period of time papers disclosing an invention and signed by the inventor as evidence of the date it was first thought of and recorded.

These disclosure documents may be forwarded to the Patent Office by the person originating the invention, by the owner of the invention (who, as you shall see, may be someone altogether different), or by the attorney or agent of the inventor or owner. Where more than one inventor is responsible for an invention, any one or all of the inventors may forward these documents to the Patent Office.

The Patent Office will retain the documents for two years, after which time the documents will be destroyed unless referred to in a separate letter submitted in connection with a related patent application filed within the two-year period. Since the documents may be destroyed, if you wish to take advantage of this service it is recommended that you retain in your possession a copy of any of the papers you are sending to the Patent Office under the Disclosure Document Program.

The disclosure document is not a patent application (a subject covered in a later chapter), and the date of its receipt in the Patent Office will not become the effective filing date of any patent application filed later on the disclosed invention. However, like patent applications, these documents will be kept in confidence by the Patent Office.

This program does not diminish the value of the conventional witnessed invention records which were referred to previously. Instead, think of it as a supplement to them.

The substance of the disclosure document submitted to the Patent Office should be approximately the same as the witnessed invention record that you have made and retained. That is, it should contain a clear and complete explanation of the invention, its construction, and mode of operation. The description of the invention should be in sufficient detail to enable a person having ordinary knowledge in the field of the invention to make and use it. When the nature of the invention permits, you should include a drawing or sketch. The use (utility) of the invention should be described where possible.

There are certain formalities that you must follow in the case of disclosure documents submitted to the Patent Office. The disclosure document must be limited to written matter or drawings on paper or other thin, flexible material, having dimensions or being folded to dimensions not to exceed 8-1/2 x 13 inches. Photographs are acceptable. Each page should be numbered. Text and drawings should be sufficiently dark to permit reproduction with commonly used office copying machines.

A nominal fee (at the time of this writing, $10) is charged by the Patent Office for this service. Since this fee may change at some future date, it is suggested that you write to the Patent Office asking for its latest fee schedule before taking advantage of the Disclosure Document Program. Information about the service can be obtained by writing to the Commissioner of Patents and Trademarks, Washington, D.C. 20231.

A check or money order for the proper fee must be submitted with the disclosure document. In addition to this fee, the disclosure document MUST BE ACCOMPANIED BY A STAMPED, SELF-ADDRESSED ENVELOPE AND A SEPARATE PAPER IN DUPLICATE, signed by you stating that you are the inventor and requesting that the material be received for processing under the Disclosure Document Program.

Your request may take the following form:

> The undersigned, being the inventor of the disclosed invention, requests that the enclosed papers be accepted under the Disclosure Document Program, and that they be preserved for a period of two years.

The papers will be stamped by the Patent Office with an identifying number as well as the date of receipt, and the duplicate request will be returned in the self-addressed envelope together with a warning notice indicating that the disclosure document may be relied upon only as evidence and that a patent application should be diligently filed if patent protection is desired.

As I've mentioned, the disclosure document will be preserved for only two years after its receipt by the Patent Office and then will be destroyed unless it is referred to in a separate letter in a related patent application filed within the two-year period. The letter filed in the related patent application must identify not only the patent application but also the disclosure document by its title, number, and date of receipt. Acknowledgment of receipt of such letters will be made in the next official communication or in a separate letter from the Patent Office.

I cannot overemphasize the fact that the two-year retention period of this program is not in any way to be considered a "grace period" during which you can wait to file your patent application without possible loss of benefits. As you've seen already, failure to file a patent application in a timely manner can destroy your ability to obtain a valid patent. The Disclosure Document Program in no way changes this.

Debunking a Few Myths

Time and again, I have heard people say that they've protected their rights in their inventions and ideas by enclosing written descriptions and drawings in self-addressed envelopes and mailing the envelopes to themselves. They feel that the date of the postmark provides them with sufficient evidence of the date they thought of the idea or invention. Occasionally, people embellish this practice by having their letters registered or certified.

The practice of sending "boomerang" letters, whether registered or not, is really a very poor way to establish proof of inventorship. First of all, envelopes can be tampered with. It is possible to open and reseal them without leaving obvious evidence of that fact. The main problem, though, is that the courts and the Patent Office rather strictly apply the rule that proof vital to an inventor's rights can only be established by corroborating testimony of competent witnesses.

No doubt many inventors are reluctant to describe their ideas to others out of fear that they will be misappropriated. Actually, the risk of theft is very minor if you choose witnesses with care. Also, of course, the fact that a person has witnessed and signed a disclosure as a witness makes it very difficult for him to allege later that he, rather than the party who prepared the disclosure, is, in fact, the originator of the idea. Since, if witnesses are chosen carefully, the chances of loss of an idea by theft are very small, I think it is far preferable to take this minimal risk than try to prove conception of an idea by sole reliance on a postmark, evidence given very little, if any, weight by the courts and Patent Office.

It is the practice of some inventors to have their written disclosures notarized in addition to being witnessed. While there is certainly nothing wrong with this, such a formality really does not add much credibility either. A notarization merely attests to the fact that the description has, in fact, been signed by you and the witnesses at the time of the notarization. No other legal significance arises by application of the notary's seal and signature, and testimony of the persons who signed the document will be needed to establish the conception date whether the document is notarized or not. A notary can, of course, be used as one of the witnesses if he has in fact read the disclosure and understood it. The fact that a notary is a notary, though, does not render him a more credible witness than any other witness.

A FINAL WORD ABOUT CONCEPTION OF AN INVENTION

As indicated above, the Patent Office awards patents to the "first inventor." When making a determination as to who is, in fact, the first inventor, the Patent Office and courts consider not only the date that the inventor thought of the idea (that is, the conception date), but also the date that a machine or article embodying the invention was actually built or the date that a patent application was filed—that is, the date the invention was reduced to practice. The inventor who proves to be the first to conceive the invention AND the first to reduce it to practice will be held to be the first inventor. An inventor who has thought of an approach is, therefore, only halfway home, and should either proceed with the actual construction of a working embodiment of the invention or consider the possibility of filing a patent application at the earliest possible date.

5

WHAT IN THE WORLD?
THE PATENTABILITY SEARCH

WHAT IN THE WORLD?
THE PATENTABILITY SEARCH

WHY A SEARCH

You cannot obtain a valid patent on something that is not "new," as that term is defined in the patent laws. Among other things, a development isn't "new" if it was patented or described in a printed publication in the U.S. or elsewhere prior to its invention by a patent applicant. Before going to the bother and expense of preparing and filing a patent application on your invention you will find it advisable to search the Patent Office files of U.S. and foreign patents and other publications to obtain at least some assurance that you are not barred from obtaining patent protection on these grounds. This procedure is called a patentability search.

I recommend a search even if you think you have an extensive knowledge of the field of your invention and believe that your concept is unique. It is more than frustrating to find, after you have invested a good deal of money and time in a patent application, that a "dead ringer" of your invention is shown in a publication or in an earlier patent issued to someone else. Just because you haven't seen something around doesn't mean that it hasn't been thought of. You'd be surprised how often a supposedly "new" idea turns up in the patent literature even though it's never seen the light of day commercially.

The results of a search also provide guidance concerning preparation of your patent application. Your attorney must know how your particular invention differs from what already exists so that he or she can prepare a worthwhile application. You want an application to be of the broadest possible scope under the circumstances. How can this be accomplished if you don't know the most important circumstance— just how unique your invention is?

CONDUCTING A SEARCH IN THE PATENT OFFICE

Are you ready for some truly mind-boggling figures? The U.S. Patent and Trademark Office has on file in its Scientific Library at Crystal Plaza, 2021 Jefferson Davis Highway, Arlington, Virginia, a compilation of over 120,000 volumes of scientific and technical books in various languages. These books are available for public use as are about 90,000 bound volumes of periodicals devoted to science and technology, the official journals of 77 foreign patent organizations, and over 12 million foreign patents in bound volumes.

And that's not all. The Scientific Library includes a Public Search Room where anyone may search and examine the nearly 5 million United States patents granted since 1836. We are talking about the greatest repository of scientific and technological information in the world. Finding anything in a collection of this magnitude would be virtually impossible if there were not some arrangement for bringing order out of chaos. Fortunately, there is. All of these U.S. patents are arranged by a classification system devised by the Patent Office. The sophisticated system incorporates over 500 subject classes and 200,000 subclasses. Every patent is classified in a specific subclass and may be cross-referenced in other subclasses. Certain employees in the Patent Office devote all their working time to fine-tuning and updating the classification system.

Locating patents in a field of technology or science covered by the Patent Office Classification System requires a good measure of judgment and a working knowledge of three principal tools that the Patent Office has provided for locating patents in a specific area of interest. These tools, available for use in the Public Search Room, are:

1. The INDEX to U.S. Patent Classification
2. The MANUAL of Classification
3. The Class and Subclass DEFINITIONS

The Index is an alphabetical list of subject headings followed by the numbers of specific pertinent classes and subclasses in the classification system that should be referred to when looking for patents relating to a particular subject. The Index is particularly useful as an initial means of entry into the system for people either lacking experience in the use of the classification system or unfamiliar with the particular technological field under study.

The Manual lists the numbers and descriptive titles of all classes and subclasses. Because of the sheer number of these, they are somewhat brief. You should consider them merely indicative and not necessarily definitive of what exists in any given class or subclass. The definitive statements of what each of the classes and subclasses encompass are in the Definitions, together with search notes which have been added by Patent Office classifiers to direct searchers to related subject matter in other classes and subclasses.

How these tools are used can best be illustrated by way of example. Suppose you have a new idea for a pair of goggles. Your goggles employ a special kind of headband and include an anti-glare shield that the wearer can flip over the goggle lenses.

Your first step would be to look up the subject matter you have in mind in the Index. You'd look up the most likely descriptive term. In this case you would, of course, look up "goggles."

The Index indicates that goggles are to be found in Class 2 (the apparel class), Subclass 14. You would confirm this by looking up Class 2 in the Manual.

Class 2 in the Manual, as luck would have it, not only lists the broad topic "goggles," but provides further breakdowns of that general subject into a variety of features including anti-glare shields and headbands, the very things of particular interest to you.

Some subclasses are numerical only. Others are alphanumeric—combinations of both numbers and letters. An alphabetical designation following the numeric designation means that the subclass has been still further broken down by the patent examiners (not the official Patent Office classifiers) into subclasses (called unofficial subclasses) relating to even more specific features to facilitate searches within the areas under their jurisdiction.

It is important to note that the Public Search Room contains only those patents listed according to the official (numeric) classification; if you were to conduct your search in the Public Search Room files, you would not find these unofficial (alphanumeric) subclasses.

The unofficial subclass files are to be found where the examiners work, in the examining groups themselves. If you wish to avail yourself of the unofficial subclass breakdown, you will have to go to the examining groups' offices located in the same complex of buildings as the Public Search Room. Actually, this is not a bad idea in any event since most patent examiners will be happy to help identify for you all of the classes and subclasses that should be covered to do a complete job. Locating the correct examining group and its location in the Patent Office is quite easy since the Manual identifies for you the particular examining group which handles the classes and subclasses you are interested in.

After you have checked out the most obvious classes and subclasses for your invention, you will want to refer to the class and subclass Definitions to see if there are any other related classes and subclasses that should be looked at to ensure a thorough and complete search of the prior art patents. "Prior art" is one of those terms you are likely to come across now and then when you work with patents. It simply means what already exists in a particular field.

As mentioned above, making a proper search calls for a good deal of judgment. It is not a purely mechanical thing, and a searcher must exercise a certain degree of imagination to ensure that all classes and subclasses which might show the specific features of the invention have been covered. For example, if your goggle shield employs a special kind of hinge felt to be new, you will want to extend your search into the class for hinges as well as into other classes and subclasses, which may be totally unrelated to goggles, wherein a similar hinge construction is likely to be found. These classes and subclasses may or may not be included in the Definitions search notes. You see what I mean about the need for experience and expertise.

If you have any doubts at all about whether you have hit all of the classes and subclasses pertinent to your invention and its features, you should request assistance from either the Public Search Room employees or the patent examiners themselves. It has been my experience that these dedicated public employees are most happy to share their expertise with people who request assistance. To assure a thorough search, there is no substitute for a good intuitive feel for the classification system based on years of experience and a sound knowledge of the system.

The Index and Manual may also be purchased from the Superintendent of Documents, U.S. Government Printing Office, Washington, D.C. 20402. Copies of the Definitions—class by class—may be purchased from the Patent and Trademark Office, Information Resources Branch, Washington, D.C. 20231.

Incidentally, to clear up a point that may be puzzling you, the U.S. Patent and Trademark Office is physically located in Arlington, Virginia, in a modern complex of buildings on the Jefferson Davis Highway even though a Washington, D.C., mailing address is used. The Public Search Room in Arlington is open from 8:00 A.M. to 8:00 P.M., Monday through Friday, except on legal holidays.

Searching Outside the Washington, D.C., Area

Not all of us live in the Washington, D.C., area or find it easy to travel there, so you are probably asking yourself how to conduct a search without actually going to the Patent Office.

Lists of patents contained in the specific subclasses of the field of search you have identified may be ordered from the Patent Office. Printed copies of the patents then may be inspected in various libraries around the country, called Patent Depository Libraries, that maintain numerically arranged collections of U.S. patents. The scope of these collections. varies from library to library, ranging from patents of only recent years to all of the patents issued since 1790.

These patent collections (organized in patent number sequence) are open to public use. Each of the Patent Depository Libraries, in addition, offers the publications of the U.S. Patent Classification System (e.g., the Manual, the Index, the Definitions, etc.) as well as other patent documents and forms. A technical staff is on hand to provide assistance.

Available in all Patent Depository Libraries is the Classification and Search Support Information System (CASSIS), an on-line computer data base. With various modes, it permits identification of appropriate classifications to search, provides numbers of patents assigned to a particular classification to permit finding the patents in a numerical file of patents, gives the current classification of all patents, and permits word searching on classification titles, abstracts, and the Index.

Facilities are generally provided for making paper

copies either from microfilm in reader-printers or from bound volumes.

Due to variations in the scope of patent collections among the Patent Depository Libraries and in their hours, you probably should call before visiting a particular library.

The Patent Depository Libraries are:

STATE	NAME OF LIBRARY
Alabama	Auburn University Libraries Birmingham Public Library
Alaska	Anchorage Municipal Libraries
Arizona	Tempe: Noble Library, Arizona State University
Arkansas	Little Rock: Arkansas State Library
California	Irvine: University of California, Irvine Library Los Angeles Public Library Sacramento: California State Library San Diego Public Library Sunnyvale: Patent Information Clearing House
Colorado	Denver Public Library
Delaware	Newark: University of Delaware
Florida	Fort Lauderdale: Broward County Main Library Miami-Dade Public Library
Georgia	Atlanta: Price Gilbert Memorial Library, Georgia Institute of Technology
Idaho	Moscow: University of Idaho Library

Illinois	Chicago Public Library
	Springfield: Illinois State Library
Indiana	Indianapolis-Marion County Public Library
Louisiana	Baton Rouge: Troy H. Middleton Library, Louisiana State University
Maryland	College Park: Engineering and Physical Sciences Library, University of Maryland
Massachusetts	Amherst: Physical Sciences Library, University of Massachusetts
	Boston Public Library
Michigan	Ann Arbor: Engineering Transportation Library, University of Michigan
	Detroit Public Library
Minnesota	Minneapolis Public Library and Information Center
Missouri	Kansas City: Linda Hall Library
	St. Louis Public Library
Montana	Butte: Montana College of Mineral Science and Technology Library
Nebraska	Lincoln: University of Nebraska-Lincoln, Engineering Library
Nevada	Reno: University of Nevada Library
New Hampshire	Durham: University of New Hampshire Library
New Jersey	Newark Public Library
New Mexico	Albuquerque: University of New Mexico Library

New York	Albany: New York State Library Buffalo and Erie County Library New York Public Library (The Research Libraries)
North Carolina	Raleigh: D.H. Hill Library, North Carolina State University
Ohio	Cincinnati & Hamilton County Public Library Cleveland Public Library Columbus: Ohio State University Libraries Toledo/Lucas County Public Library
Oklahoma	Stillwater: Oklahoma State University Library
Oregon	Salem: Oregon State Library
Pennsylvania	Cambridge Springs: Alliance College Library Philadelphia: Franklin Institute Library Pittsburgh: Carnegie Library of Pittsburgh University Park: Pattee Library, Pennsylvania State University
Rhode Island	Providence Public Library
South Carolina	Charleston: Medical University of South Carolina
Tennessee	Memphis & Shelby County Public Library and Information Center Nashville: Vanderbilt University Library
Texas	Austin: McKinney Engineering Library, University of Texas College Station: Sterling C. Evans Library, Texas A & M University Dallas Public Library Houston: The Fondren Library, Rice University

Utah	Salt Lake City: Marriott Library, University of Utah
Washington	Seattle: Engineering Library, University of Washington
Wisconsin	Madison: Kurt F. Wendt Engineering Library, University of Wisconsin Milwaukee Public Library

SEARCH SERVICES

You *can* do your own patentability search, but do you really *want* to? After all, the process will be quite lengthy and tedious if you aren't familiar with the process. Then, too, the final results provided by an inexperienced searcher are not likely to be on par with those produced by someone who does searches for a living—on a day-to-day basis. And don't forget about the hassle involved in just getting to a place where you can make a search. Although Patent Depository Libraries are fairly accessible across the country, they are not exactly located on every other street corner. For these reasons, the vast majority of inventors choose to hire someone else to conduct a search.

There are two main ways to go about this. Let me give you my preferred approach first. I recommend that you make arrangements through your patent attorney for a patentability search. Every patent attorney has a working relationship with a professional searcher in the Washington, D.C., area with direct access to Patent Office files and records—someone who knows the patent classification system like the back of his hand. You will get a prompt, decidedly thorough job done when your attorney retains the services of a professional searcher. In any patentability search, the most important thing is the end result. You want to end up with a collection of the prior art closest to your invention. You don't want to be surprised by a close reference cited by the Patent Office examiner that you weren't aware of before filing your application. Your patent attorney knows who can get these results because he or she works with them on a regular basis. Before you request your patent attorney to arrange for a search, it is a good idea to ask for an estimate of the costs. Search costs vary among different practitioners and also vary depending upon the complexity of the invention to be searched. Also, the search fee will depend somewhat on whether you wish to have the search cover only U.S. patents or both U.S. and foreign patents, and possibly non-patent literature such as trade journals.

Most inventors prefer to save expenses by omitting the search of foreign patents and non-patent literature, although there is, of course, the outside possibility that an invention may be disclosed in a foreign patent or in non-patent literature and not in a U.S. patent. Your patent attorney can provide you with guidance about the likelihood of this occurring.

One way to go about things is to establish an initial reasonable limit on search expenditures. For example, you can tell your attorney to instruct the searcher to stop when he hits a certain billing charge, say, $200 for a very simple invention. Your attorney and you can then look over the prior art located in the search and decide whether it is worthwhile to invest another $100 or so. A good searcher has a real intuitive feel for these things and can tell you whether or not the search up to that point has hit all of the likely classifications for an invention like yours.

ADVERTISEMENTS OFFERING SEARCH SERVICES

Instead of making arrangements through your attorney to have a search made, you can contact a searcher directly, but be careful. A number of the popular magazines aimed at the "do-it-yourselfer" or science buff contain advertisements by firms or individuals offering low cost searches. Sometimes the costs quoted for a search are as low as $30 or $40. BEWARE. Anyone taking advantage of an offer of this nature gets precisely what he pays for—at best, a superficial search, the results of which will be of little value in assessing the merits of an invention. A slap-dash search can be quite harmful since it can lead you to believe that your invention is new when, in fact, it is not. As you've seen, a thorough search is often a complex and time-consuming business calling for expert skills. It stands to reason, doesn't it, that you are unlikely to obtain the benefit of such skills for a mere pittance.

Locating a competent professional who can do a good job of searching is not easy, and I think you'd be well advised to steer clear of all advertisements placed by searchers or, for that matter, those searchers listed in the Yellow Pages of telephone directories unless they are recommended by someone you trust. The Patent Office will not endorse or recommend specific searchers, and it has almost no control over companies and individuals offering search services. Virtually anyone can say that he or she is a patent searcher. I'm not saying that there aren't good searchers out there whom you can contact on your own. There are. But the question is—who are they? Picking a name at random out of the Yellow Pages or a magazine classified ad section is a risky way to go, if you ask me.

And, don't forget, if you use a patent attorney to arrange for a search, he or she will be able to provide you with an expert analysis of the search results, with particular attention being given to the question of how the located patents affect your chances of obtaining your own patent. A fee will be charged for this service based on the patent attorney's hourly billing rate. You should ask for an estimate. The analysis for the simplest of inventions could easily run $200 or more. Of course, the fee owed to the person who actually gathered the information by conducting the patent search will be added to the attorney's fee for analysis. Searchers who are not patent attorneys are severely limited by law insofar as giving advice of this nature is concerned. In fact, they cannot legally give you an opinion as to whether your invention is patentable or not.

6

Everything You Wanted to Know about a Patent and Were Afraid to Ask

EVERYTHING YOU WANTED TO KNOW ABOUT A PATENT AND WERE AFRAID TO ASK

THE U.S. PATENT SYSTEM

Our patent system has been around a long time and it has served the nation well. A bill establishing the system was passed by Congress and signed into law by President George Washington in 1790. Thomas Jefferson had responsibility for its administration, and the first patent was granted on July 31, 1790. The document is an autograph hound's delight. It was signed by George Washington, President of the United States; Edmund Randolph, Attorney General; and Thomas Jefferson, Secretary of State.

Many changes have taken place in the patent system since that time, as they have in the country itself. In spite of these changes, the patent system continues to do what it has always done—encourage invention and technological advancement by providing an incentive to the inventor. The incentive is in the form of an exclusive right in the invention for a limited period of time so that the inventor can reap its financial rewards. In effect, a patent grants a limited monopoly.

As Abraham Lincoln expressed it: "The patent system added the fuel of interest to the fire of genius." Honest Abe, as usual, knew what he was talking about. Years before being elected President, he was issued his own patent on May 22, 1849, for a "Device for Buoying Vessels over Shoals." As was the required practice in those days, a model of this invention, assembled by Lincoln's own hands, was submitted to the Patent Office. The model can now be seen in the Smithsonian Institution in Washington, D.C.

The basic patent law, as it exists today, came into effect January 1, 1953. It specifies the types of things for which patents can be granted by the U.S. Patent and Trademark Office.

That agency is one busy place. Patent applications are now being filed at a clip approaching 150,000 a year. Approximately 1,400 examiners give them a thorough going over and, when the examining process is completed, about 65 percent of the patent applications are issued as patents.

WHAT IS A PATENT?

Getting a bit more specific, a patent is a grant issued by the federal government giving an inventor the right to exclude all others from making, using, or selling his or her invention within the United States, its territories, and possessions. Patents are granted for a term of 17 years, subject to payment of maintenance fees. Design patents, that is, patents issued on ornamental designs, are exceptions to this general rule and are for 14 years.

People often confuse patents with copyrights and trademarks. These are totally different things. Although I'll be covering copyrights and trademarks in more depth a little later on, it's worthwhile at this point to discuss a few of the major differences.

A copyright does not protect inventions, and the Patent Office has nothing at all to do with copyrights.

A copyright protects the creator of certain types of literary and artistic works, such as books, paintings, musical compositions, photographs, motion pictures, and sculptural works of art, from having those works copied by someone else without permission.

A description of a machine, for example, could be copyrighted as a writing, but this would only prevent others from copying the description, not from writing this description independently on their own nor from making and using the machine described. On the other hand, a patent would prevent the copying of the machine itself. Copyrights are registered in the Copyright Office in the Library of Congress, an agency quite independent from the Patent Office.

Another form of intangible property that causes some confusion is the trademark. A trademark is a word or design, or a combination of both, used by someone to distinguish his or her products or services from those of others. Trademark rights are used to prevent someone else from using similar marks which may cause customer confusion. Trademark rights arise under common law (nonstatutory law) and, in addition, trademark registrations are granted by both state governments and the federal government. Federal trademark registrations are granted by the U.S. Patent and Trademark Office, although this operation of the office is separate and distinct from the one responsible for issuing patents.

SHOULD YOU APPLY FOR A PATENT?

As you will see later on, the patent statute is quite specific as to the types of things that can be patented.

Aside from this question, however, even if something is patentable, other factors may make it inadvisable to spend the time and considerable expense preparing and filing a patent application.

A patent is by no means a guaranteed road to riches. A profit can be obtained from a patent only if it covers something that people are willing to pay for. Your invention should be an improvement over other known approaches and provide advantages if you expect to make money from it.

This is where your business judgment comes in. Your patent attorney can provide input, but the final call is yours. To file or not to file—that is the question awaiting your reply.

Here is what you want from your attorney. Have your attorney identify for you the specific features of your development that he or she feels are patentable. Then get an estimate of the costs of preparing and filing a patent application (or applications) on these features.

Preparation and filing of a patent application can run on the order of $2,000 or more for a relatively simple invention. Complex inventions will be much costlier since the patent lawyer's fee will for the most part be a function of the time he or she will have to spend on it. Costs also include the government filing fee and other out-of-pocket expenses such as the cost of drawings included in the application.

Now you have the ball. Make a realistic assessment of what the patentable features might be worth to somebody. Are they something that would not really add much in the way of value, or do they have real significance? Can they save money? Do they have operational or other commercial advantages over what already exists?

The results of the patentability search will probably have shrunk your preconceived notion of what you have invented and the contribution it can make to the marketplace. You may find that you no longer have a breakthrough invention, just a mere shadow of one. Instead of coming up with a radically new concept, you may have found that you have reinvented the wheel and that your only contribution is a minor fitting on the wheel's spoke.

Is that fitting worth a patent application? How do the potential rewards now weigh against the costs of application preparation and filing?

Bear in mind that the narrower your invention, the greater the options available to someone who wishes to skirt around a patent directed to that invention. If you've invented the wheel, you own the world. Spoke fittings, though—well maybe there are hundreds of other types already available. You must realize that you cannot use your patent to prevent someone else from using previously known alternatives to your approach.

WHAT CAN BE PATENTED?

You're going to have to forgive me because I'm going to get into some legal talk. Patents are creatures of statutes and to understand patents properly, what they are and what they do, there is simply no way to ignore the federal laws on the subject. I'll try not to make your eyes glaze over. If they do, believe me, it's only a temporary condition.

The patent laws state that only "inventions" can

be patented, and they identify general fields within which these inventions must lie. In the words of the pertinent statute (35 United States Code 101), a patent may be obtained by whoever "invents or discovers any new and useful process, machine, manufacture, or composition of matter, or any new and useful improvement thereof." The items in this short laundry list are commonly referred to as statutory classes of invention, and anything falling outside can't be patented. For example, patent protection cannot be obtained on ideas or suggestions, methods of doing business, combinations of pure mental steps, or mere printed matter that is not associated with some mechanical feature.

While your patent attorney will let you know whether or not your invention falls within the statutory classes, you should at least be aware of their general scope.

The term "process" means a method of doing things. There are, for example, whole Patent Office classes devoted to processes, such as metalworking, refining, methods of manufacture and construction, electrical processes, and myriad other methods.

The word "machine" really causes very little difficulty since its meaning is rather obvious, covering the most complex mechanisms to those having few, or even no, moving parts.

The term "manufacture" really overlaps "machine" since a "manufacture" encompasses all manufactured articles. Some writers claim that a "machine" is a mechanism having some moving parts and that this is not necessarily the case with a "manufacture." Such discussions are rather academic, however, since you as an inventor need only be concerned with whether or not your invention falls within any one of the four statutory definitions. You need not identify the specific class that will be assigned by the Patent Office.

The term "composition of matter" refers to mixtures of ingredients and new chemical compounds. In the case of mixtures of ingredients, such as medicines, a patent cannot be granted unless there is more to the mixture than the mere effect of its components. For this reason, the so-called and misnamed "patent medicines" are more often than not unpatentable and do not fall within the statutory classes of definition since they merely employ combinations of well-known ingredients that act in the same manner in these compositions that they do when they are used separately.

I am sure you are aware, for example, that aspirin is employed as the primary ingredient in many cold remedies, and, since aspirin operates in these remedies as it normally does and has the same effect as it has in its separate form, there is no patentable result.

An Invention Must Be New to Be Patentable

In addition to the requirement that an invention fall within one of the statutory classes to be patentable, it must also be "new" or novel as defined in the patent statute. According to that statute, and I quote, an invention cannot be patented if:

> (a) the invention was known or used by others in this country, or patented or described in a printed publication in this or a foreign country, before the invention thereof by the applicant for patent, or

(b) the invention was patented or described in a printed publication in this or a foreign country or in public use or on sale in this country more than one year prior to the date of the application for a patent in the United States.

Thus, you cannot obtain a patent if the invention has been described in a printed publication anywhere in the world or if it has been in public use or on sale in this country before the date you made your invention. Similarly, if the invention has been described in a printed publication anywhere, or has been in public use or on sale in this country more than one year before the date on which an application for patent is filed in this country, a valid patent is unobtainable.

THE ONE-YEAR RULE

These statutory bars arise even when the *inventor* describes his invention in a printed publication, uses the invention publicly, or places it on sale. This is the "one-year rule" which says that an inventor must apply for a patent before one year has gone by after any one of the three foregoing activities takes place. Unfortunately, many inexperienced inventors have learned of this "one-year rule" too late. Insofar as the filing of patent applications is concerned, it is often delay, rather than haste, that makes waste.

BEING "DIFFERENT" ISN'T ENOUGH

An invention, to be patentable, must be something more than merely different from what is already known. As the patent statutes put it, a patent may not be obtained "if the differences between the subject matter sought to be patented and the prior art are such that the subject matter as a whole would have been obvious at the time the invention was made to a person having ordinary skill in the art to which said subject matter pertains." In other words, small differences or advances obvious to someone having ordinary skill in the art are not considered patentable. You can be certain that this clause has been argued about vociferously in many lawsuits.

A Patentable Invention Must Be New and Unobvious

Generally speaking, an invention, to be patentable, must give new and unobvious results as compared with known approaches. Examples of differences usually considered minor and not patentable are changes in degree or size of parts, changes of locations of parts, making a device portable, and omission of parts.

It is often said that a patentable invention must be a sum greater than the total of its parts. By that it is meant that it is not patentably inventive merely to add up a bunch of old parts or elements, be they mechanical elements, ingredients, or steps in a process, if nothing new is achieved and these elements function in the same way that they do in other combinations in which they have been used.

To Be Patentable an Invention Must Be Useful

To be patentable an invention has to be useful, or, as the courts sometimes put it, the invention must have utility. This requirement presents little difficulty since almost anything meets this standard. An invention may be patentable even though it is inferior and perhaps operates less efficiently than similar things already available. A patentable invention may still be a "loser" from a commercial standpoint. The Patent Office doesn't prejudge the chances of an invention for commercial success, and this factor has no bearing whatsoever on whether or not a patent is issued. About the only inventions not considered useful and thus not patentable are those felt to be frivolous, immoral, or designed for a fraudulent purpose.

Somewhat related to the subject of utility is the fact that an invention must be operative before the Patent Office will issue a patent on it. I indicated earlier that many years ago the Patent Office required inventors to submit working models so they could see that the inventions were in fact operable. This is no longer the case, and the Patent Office will require a working model only in a very unusual situation where it feels a serious question exists as to operability. For example, a model will be required by the Patent Office if the invention relates to a perpetual motion machine, on the theory that such a device defies established laws of physics.

An invention must, though, work on paper, as it is described in the application. If your patent attorney has any questions concerning the operability of your invention, he or she will ask you for more information so that it can be properly described in an application.

Who May Apply for a Patent?

In the United States only the inventor may apply for a patent, subject to a few exceptions. There are no restrictions as to the types of persons who can obtain a U.S. patent if they are in fact the inventors. A foreign inventor can obtain a U.S. patent under the same conditions as a United States citizen. In fact, almost half of U.S. patents are now going to foreigners.

A person applying for a patent who knows he is not the inventor is subject to criminal penalties. In addition, a patent issued to such a person would be invalid and unenforceable.

If several people work together to make an invention, the application for a patent should be applied for in the names of all co-inventors. It is necessary to dig into the facts to determine precisely which persons are in fact inventors. To be a co-inventor, a person should have contributed something toward the inventive concept. If, for example, one person provided all of the ideas and someone else has only followed that person's instructions in constructing a model, the idea contributor is the sole inventor and the patent application and patent should be in his or her name alone.

The inventor is in effect the person who contributes the ideas relating to an invention. An employer or someone merely furnishing money for building and testing the invention is not entitled to be listed as a co-inventor in the application.

As you will see later, it is possible for an inventor to sell or otherwise transfer rights in a patent or patent application to someone else by means of a properly worded assignment. An assignment may transfer all or any part of the inventor's interest in the application or patent depending upon how it is worded. The invention still must be filed in the Patent Office in the name of the true inventor, however, and it is not considered the invention of the person who has purchased the rights to the patent application.

SPECIAL SITUATIONS

A special situation exists when an inventor refuses to execute an application for a patent or cannot be found or reached after diligent effort. A person to whom the inventor has assigned or agreed in writing to assign the invention or who otherwise shows sufficient proprietary interest in the matter can apply for a patent on behalf of and as agent for the inventor. Proof of the pertinent facts must be submitted and a showing must be made that the action is necessary to preserve the rights of the parties to prevent irreparable damage. It occasionally happens, for example, that a disgruntled ex-employee harbors so much resentment that he or she refuses to sign a patent application for an invention which was contracted to be assigned to a former employer.

If the inventor is dead, an application for patent may be made by legal representatives; that is, by the administrator or executor of the inventor's estate. If the inventor is insane, the application for patent may be made by a guardian.

SOME POTENTIAL PITFALLS

A couple of basic points concerning inventorship and ownership of patent rights occasionally cause difficulties for people newly exposed to the patent system. First of all, to legally qualify as an inventor and be entitled to file a patent application in your name, you must actually have thought of the idea. You must be the true inventor. Occasionally, people return from trips out of the country with ideas that they have read about or seen during their travels. Some try to file patent applications in their own names in the U.S.

based on the mistaken belief that they are the inventors because the ideas are new to this country. Unfortunately, it is not possible to patent an idea unless it originates with you. Merely importing someone else's idea doesn't make you an inventor.

Another point that causes confusion relates to ownership. Co-inventors are joint owners of a patent if they have not assigned all of their rights. Co-inventors, or other persons owning a patent jointly, can independently grant licenses to a third person or company to make, use, and sell the invention without the consent of the other co-inventors or co-owners. Also, a joint owner or co-inventor need not obtain the consent of the other people owning a patent jointly to make, use, or sell the invention independently. This is true even though the joint owner granting a license owns only a very small percentage of the patent. Unless you want to grant this power to a person to whom you assign a part interest, you should ask your patent attorney to include special preventive language in the assignment.

As you might expect, there are occasional fallings out between joint owners and between co-inventors; the fact that a co-inventor or joint owner, in the absence of an agreement to the contrary, may independently license the invention can often work to the detriment of the other parties. Consequently, it is best to consider this matter early on. If you are a co-inventor, you will wish to protect yourself against the possibility that another co-inventor, without your approval, will license a third party under your patent. An agreement protecting all co-inventors from this possibility should be prepared by your patent attorney at a very early stage. Similarly, unless you want to grant unrestricted licensing power to a person assigned a part interest in your invention, you should include special language in the assignment to prevent this result.

APPLYING FOR A PATENT

A patent is applied for by sending to the U.S. Patent and Trademark Office a formal written application describing your invention and petitioning the Commissioner of Patents and Trademarks to grant you a patent. This procedure is called "filing" a patent application. (See the appendix on page 231 for an example.)

Under the law, you have the right to prepare your own patent application and file it in the Patent Office without the help of any attorney or agent. In addition, you are legally entitled to prosecute the application in the Patent Office without professional help. The word "prosecute" refers to the writing of letters and submission of legal arguments to the Patent Office (1) to convince the Patent Office examiner that a patent should be issued and (2) to fix the legal scope of the patent protection granted.

While you can avoid paying for the professional services of a patent attorney by preparing, filing, and prosecuting your own patent application, should you do these things on your own? I recommend against it. The Patent Office recommends against it. While there are exceptions to every rule, an inventor handling these matters on his or her own is unlikely to obtain a good patent, that is, one providing adequate protection for the invention.

What an Application Consists Of

An application for a utility patent—the most common type—includes a *specification* describing the invention. One or more *claims* are also included. In effect, a claim is a statement which spells out what you regard to be your invention. Usually more than one claim is included in an application, each artfully worded so as to be directed to different features of the invention and to define it broadly and specifically.

The application also includes *drawings* in those cases where a drawing is possible. Some inventions can be illustrated; some can't. A patent application additionally includes a signed *oath or declaration*, and, usually, a signed *power of attorney* appointing a registered attorney or agent to represent the inventor before the Patent and Trademark Office.

The power of attorney and oath or declaration are usually in very much the same form in every application submitted to the Patent Office. For this reason, these documents are often collectively called the "formal papers." A proper *filing fee* must also be submitted (see the later section on fees).

The application must be in the English language and it must be clearly written or printed in permanent ink on one side of the application papers. The Patent and Trademark Office prefers typewriting on legal-size paper, 8 to 8-1/2 x 10-1/2 to 13 inches, 1-1/2 or double spaced, with margins 1 inch on the left-hand side and at the top. The Patent Office may require replacement by typewritten or printed papers if the papers originally filed are not correctly, legibly, and clearly written.

All required parts of an application for patent must be received by the Patent Office before the application will be accepted and placed in the files for examination. The Office requires that all parts of the application be in compliance with its rules but will waive temporarily certain minor mistakes in the application, subject to later correction.

The Patent Office will notify your attorney if the papers and parts of the application are incomplete or are so defective that they cannot be accepted as a complete application for examination. A time period will be given to remedy the differences. After the allotted time, if the problem isn't corrected, the application will be returned or disposed of by the Patent Office. The filing fee may be refunded when an incomplete application is refused acceptance, but you might be charged a handling fee.

The Patent Office numbers all applications in the order they are received, and you (through your patent attorney) will be informed of the serial number and filing date of the complete application. The filing date of an application is the date on which a specification (including claims) and any required drawings are received by the Patent Office. If application papers were previously incomplete or defective, the filing date is the date on which the last part completing the application is received by the Patent Office.

OATH OR DECLARATION

The patent statutes require that you make an oath or declaration that you believe yourself to be the original and first inventor of the subject matter of your application. In addition, you must make a number of other allegations required by the statutes and by Patent and Trademark Office rules.

If an oath is used, it must be sworn before a notary public or other officer authorized to administer oaths. A declaration may be used instead of an oath as part of the original application. A declaration need not be notarized.

Under no circumstances will the Patent Office return any papers of a complete application, nor will the filing fee be returned. If you haven't kept copies, the Patent Office will furnish them for a fee. This will normally not be required since your patent attorney will customarily provide you with copies of all papers submitted to the Patent Office upon request.

Filing Fees

These fees are subject to change, but, as of this writing, the filing fee for an application for an original patent, except in design and plant cases (discussed at the end of this chapter), consists of a basic fee and additional fees. The basic fee is $340. This entitles the inventor to present 20 claims. Additional fees may be required depending upon the number and form of the claims. If the owner of the invention is considered a "small entity," the filing fees will be halved. An individual is considered a small entity and thus would only have to pay a $170 basic fee. Small businesses and nonprofit institutions are entitled to the same break.

Specification—Description of the Invention

A patent specification must provide a description of the invention adequate to teach a person skilled in the field of the invention to make and use it. A proper description is couched in full, clear, concise terms. If the invention is shown in drawings, the description must be cross-referenced to them.

Your invention should be described in the specification in such a way as to distinguish it from other inventions and from what you know to be old. It must describe in a complete manner a specific form of the invention, the way it operates, and the principles involved. The best way to carry out your invention must also be described.

Obviously, a good degree of common sense must be employed by you and your patent attorney to determine exactly what is necessary to explain your invention to that hypothetical person skilled in the art to enable manufacture and use of the invention. For example, where the invention employs well-known parts or principles of operation as part of its function, it will not be necessary to belabor the obvious. On the other hand, unanswered questions should never be raised by the specification that you as the inventor provided.

When the invention is an improvement of what is already known, the specification description should be confined to that specific improvement and to any other parts which necessarily cooperate with it or are necessary to a complete understanding or description of it. For example, if the invention relates to an automobile tie rod, it is not necessary—or proper—to describe all of the elements of a complete automobile. You only have to describe those other car parts which necessarily cooperate with the tie rod and help to point out its inventive nature.

CLAIMS

The specification has to conclude with one or more claims. The claims are really the most important part of the patent application since they define the bounds of an inventor's patent rights. They fix the scope of protection of the patent.

Skillful drafting of these claims is so important, it's a job for a professional. If an invention is entitled to broad protection based on the known prior art, it is important to draft the claims with that broad scope. However, the claims must distinguish the invention from others in existence. The most important challenge to your patent attorney will be to write claims that distinguish your invention from the prior art, yet secure coverage broad enough to give the ultimately issued patent its best chance for commercial success.

The claims must go to the essentials of the invention, and they should not include or be limited by unimportant incidental features. If they are, others may be able to use the important features without paying you anything, merely by making simple changes which eliminate the unimportant features that may have been claimed.

So that a patent attorney can do a proper job, it is very important that you bring to his or her attention all background material and information relating to the invention in your possession. For example, you should describe all pertinent prior art that you know about. In addition, you should explain to your patent attorney all of the various forms your invention might take, not merely the best one, so that the invention can be claimed in language broad enough to include all of these various alternatives.

You should also make certain that the patent attorney has in his or her possession all of the pertinent dates relating to conception and possible reduction to practice of the invention as well as the dates on which the invention was described in a printed publication, placed on sale, or available for public use. That way work on the application can be paced and scheduled to ensure filing in the Patent Office before a statute date could possibly prevent that. Remember the one-year rule?

Claims must conform to the invention as described in the specification description. Terms and phrases used in the claims must also be used in the description so that their meanings in the claims can be readily understood by reference to the description.

In summary, claims are really brief descriptions of the invention in which unnecessary details have been eliminated and in which all essential features necessary to distinguish the invention from that which is old are recited. Claims are the real guts of the patent, and any questions concerning novelty, patentability, or infringement of the ultimately issued patent are judged on the basis of the claims.

THE DRAWINGS

Whenever the nature of the case permits it, and that's most of the time, an applicant for a patent must furnish drawings of the invention.

The drawings must show every feature of the invention as defined in the claims and, in addition to illustrating the preferred form of the invention, must show alternative embodiments of the invention if their features are claimed.

The Patent Office requires that drawings be in a particular form and meet certain requirements. The size of the sheets on which the drawings are made is specified, as are the type of paper, margins, and other details relating to the nature of the drawings themselves. The reason for these standards is that uniformity is required for printing and publishing the drawings when the patent issues. It almost goes without saying that the drawings must be clear and readily understandable by people reading and using the patent description.

All patent attorneys are familiar with the Patent and Trademark Office requirements for drawings. Typically, the patent attorney will make arrangements with a skilled draftsman to do the drawings in accordance with the attorney's instructions. Even if you are a skilled technical illustrator, I recommend not doing the drawings yourself because of the special technical rules applicable to them. If you would like more information about the exact nature of these rules, you can obtain the booklet "Guide for Patent Draftsmen" by writing to the Patent and Trademark Office, Washington, D.C. 20231.

Drawings not complying with all of the regulations of the Patent Office will be accepted for purposes of examination, but correction or new drawings will be required later.

How You Can Help Your Attorney—and Save Money

Most patent attorneys charge by the hour and these hourly rates commonly hover above the century mark. It doesn't take too much time at $100 per hour or more to add up to some real money. It stands to reason that anything you can do at your end to save your attorney's time will pay handsome dividends.

What can you do? First and foremost, give your attorney as much information and detail concerning your invention as you possibly can. Present him or her with a complete description in writing. Describe the *best* form of the invention, as well as other possible forms. If you have a model, give your attorney that. If you don't have a model, furnish sketches or drawings. These need not be fancy. They can be drawn freehand for that matter, but they should provide a clear picture of the parts of the invention, how they cooperate, and how they work. If dimensions or other specifications are critical, these should be disclosed.

The objective here is to answer questions concerning your invention—its construction and operation—before they are asked. You don't want your lawyer to engage in guesswork when the application is being drafted. You want him to go about this task efficiently and without a lot of wheel spinning. Phone calls and in-person conferences between you and your attorney to deal with these issues can quickly run into big bucks.

Abraham Lincoln once said that a lawyer's time and advice are his stock in trade. Try to maximize the advice and minimize the time as much as possible. Something you can do to this end is to provide your attorney with a statement describing how your invention is distinguished from things already in existence. What are the invention's chief advantages? What objectives are being met?

Some prolific inventors, those who have been

around the track a few times, have carried things a step further. The descriptions they give to their attorneys are virtually in the form of a patent application specification. The objective is that the attorney will do a nip here and a tuck there, put an insert here and delete something from there, so that the specification is in final form.

This almost never works out in practice and certainly would never be possible unless an inventor has lots and lots of exposure to patents and the patent process. No harm in trying this, of course, but don't count on this procedure saving much attorney time. Leave the lawyering to the lawyer. Just give your patent attorney all the background and information needed to do the job efficiently.

FILING A PATENT APPLICATION

After your application has been prepared, you should go over it very thoroughly. You (and any co-inventors) should carefully read the application and make certain you are satisfied as to its accuracy and completeness. If all is in order, you should then sign the formal papers your patent attorney will have prepared and leave the application in his hands for filing. If rights to the invention are being assigned, a document for this purpose will have been prepared and will be filed in the Patent Office along with the application.

PROCEEDINGS IN THE PATENT OFFICE

Exactly what happens after your application is filed with the Patent Office? First, the application is given a docket serial number and official filing receipt date. In the event the application for some reason is not considered complete, your attorney will be notified and given the opportunity to complete it.

After it is given a filing number and receipt date, the application is assigned for examination purposes to a particular examining group within the Patent Office and to a particular examiner within that group. The examiner then normally takes the application up for examination in the order it was received relative to other pending applications.

Occasionally, applications are taken up for examination out of order, upon request, if certain special conditions have been met. This is known as making an application special. One way an application can be made special is by submitting proof to the Patent Office that the invention covered by an application is being manufactured, used, or sold by someone else. Other grounds for making an application special do exist, and you should explore the subject with your patent attorney in greater detail if you are interested in doing this.

EXAMINATION

When your case comes up for examination, normally six months to a year after filing unless made special, the examiner will check to see that the application complies with the requirements of the patent laws and Patent Office rules. In addition, he or she will conduct a search of the prior art, mostly existing U.S. and foreign patents, to see if the invention is new. During the examination procedure, the examiner goes through each and every claim of the application, reviewing its language in light of the located prior art references. On this basis, he or she decides whether a particular claim will be allowed.

THE OFFICE ACTION

Notification of the examiner's decision is by means of an "action," a written communication usually mailed to the applicant's patent attorney. In the action, the examiner indicates which, if any, claims are considered patentable and notes any other objections or requirements. The patent attorney is provided with copies of patents or other prior art references cited by the examiner.

The action will normally be complete, covering all issues. On occasion though, the examiner finds fundamental defects in the application. In these cases, the action may be limited to these questions and examination of the claims deferred.

Only one invention may be claimed in a single application; if the Patent Office feels that two or more inventions are covered, it will require that the application be limited to only one. A separate application may be filed on the other invention; if it is filed while the first application is still pending, it will be entitled to the benefit of the first application's filing date.

In an action dealing with the actual merits of the claims, the examiner rejects any claim not considered patentable for one reason or another. The precise references relied on by the examiner will be pointed out with respect to each rejected claim.

Don't be discouraged if claims are rejected in the first action by the examiner. Relatively few applications are allowed in the form originally filed. It's not unusual for *all* claims to be rejected in the first round.

Replying to the Office Action

After receiving the Patent Office action, your patent attorney will customarily notify you and advise you of the position taken by the Patent Office examiner. You are entitled to a copy of the Patent Office action from your attorney, and if you aren't provided one, feel free to request it.

The attorney will study the action and give you his or her conclusions and recommended course of action. It is often helpful for attorney and client to have a conference after both have had an opportunity to consider the action and cited references so that a course of action may be mutually plotted.

If the action is adverse in any respect and if you wish to continue your quest for a patent, the action must be replied to within a certain time specified by the examiner, usually three months from the date of action. Within this period of time you may request reexamination or reconsideration of the application, and the application can be amended to try to overcome any rejections or objections made by the examiner.

In the examination proceeding, the usual bone of contention between an applicant and the Patent Office lies in the language of the application claims, which, of course, define the scope of the invention. Your attorney may request cancellation of rejected claims or may amend them so that they define more clearly the prior art cited by the examiner.

Of course, there is the further alternative of leaving the rejected claims in the form originally submitted and attempting to change the examiner's mind through arguments. Quite frankly, examiners are usually not prone to change their minds regarding the allowability of claims without amendments having been made, although this does happen occasionally, such as when an incorrect interpretation of a reference has been made by the examiner. It is possible to arrange a personal meeting with the examiner. These interviews are usually held between the examiner and the patent attorney, although an applicant is certainly free to sit in. Sometimes that is the smart thing to do. After all, who knows more about the invention than the inventor?

It is important to keep your patent attorney fully informed of any changes made to or planned for your invention. While the Patent Office does not permit an application to be changed to describe new matter such as improvements after it has been filed, you will want to be sure that the claim language covers these improvements, if possible. If the improvements cannot be protected by the claims of the application already on file, your patent attorney may recommend filing an additional application specifically directed to the improvements.

FINAL REJECTION

After your attorney files an amendment or other response, the Patent Office will reexamine the application, giving due consideration to the submitted arguments. If new or amended claims have been presented, they will be considered. Another action will be mailed. It may be that the examiner will find all claims allowable and the application in condition for issuance as a patent. What happens when this determination has been made will be discussed a little later. On the other hand, the examiner still may find all or some of the claims currently in the application to be unpatentable and reject them.

This second Office action is usually indicated as being final, in other words, the examiner's last word on the subject. To get the application issued as a patent at this juncture requires an amendment canceling and/or modifying claims to put all claims remaining in the case in allowable form. If a single claim remains a bone of contention between you and the examiner, an appeal from his or her decision is generally the only way to resolve matters in your favor.

Appealing a Rejection of Claims

An appeal may be taken to an appellate branch of the Patent Office called the Board of Patent Appeals if the examiner persists in rejecting an application's claims. A written brief must be submitted to the Board to support your position, and an appeal fee is required. The Board may make its determination directly from the written record, or, at your option, this may be supplemented by an oral hearing.

The Board of Patent Appeals is still not the final arbiter as to the question of rejection of claims since a further appeal to the federal court system may be taken if the decision of the Board goes against you. Bear in mind that these appeals and the lawyering associated with them can cost a great deal of money. By all means, get an advance estimate from your attorney before proceeding. It may be that the expense is worth it. Perhaps not.

Allowance of an Application and Issuance of a Patent

If a patent application is found to be allowable by the Patent Office, then payment of an issue fee must be made. If payment is not made, the application will be regarded as abandoned or forfeited.

Upon payment of the issue fee, the patent issues as soon as possible, depending upon the backlog of printing. On the day of its grant or as soon thereafter as possible, the original patent is delivered or mailed to your attorney. On the granting date, the record of the patent in the Patent Office is opened to the public and printed copies of the patent become available.

Interferences

It occasionally happens that two or more applications are filed by different inventors directed to the same invention. Under the law, a patent to an invention can only be granted to the first inventor, and it is necessary for the Patent Office to determine which of the applicants is in fact the first inventor. This determination is made in a proceeding known as an "interference."

Statistically, only about 1 percent of the applications filed become involved in proceedings of this nature. Interference proceedings may also be established between an application and a patent that has already been issued to another party, if the patent has not been issued for more than one year prior to the filing of the conflicting application and, additionally, provided that the conflicting application is not barred from being patented for some other reason.

An interference proceeding can be a drawn-out, costly business, and the potential worth of some inventions is simply not worth the time and effort that may be involved. The Patent Office encourages settlement of the question of inventorship by the parties themselves if this is possible, and you will no doubt wish to explore this possibility with your patent attorney if you have the misfortune to become involved in an interference.

In an interference proceeding, each party must submit evidence of facts proving when the invention was made. A party not submitting evidence is restricted to the date of filing of his application as the earliest date; thus, you can see the importance of keeping and maintaining written records concerning your invention.

It is in the interference proceeding that the legal concepts of "conception of the invention" and "reduction to practice" come into play. As previously discussed, conception of the invention refers to the devising of the means for accomplishing the result—the thinking of the idea—while reduction to practice refers to the actual construction of the invention in physical form. In the case of a machine, reduction to practice includes the actual building of the machine. In the case of an article or composition of matter, it includes the actual making of the article or composition. And in the case of a process, it includes the actual carrying out of the steps of the process. Actual operation, demonstration, or testing for the intended use of the invention is also usually necessary.

As you've already seen, the filing of an application for a patent completely disclosing the invention is treated as equivalent to an actual reduction to practice. Filing is referred to as a constructive reduction to practice. However, if an operating embodiment of the invention has actually been constructed and used prior to the filing of a patent application, this earlier date would be to your advantage in an interference proceeding.

The question of prior inventorship can get to be a complicated one, and it is an area calling for a considerable amount of legal expertise and familiarity with interference practice. In the most simple situation, the inventor proving himself to be the first to conceive the invention and the first to reduce it to practice will be held to be the prior inventor; but there are more complicated situations which cannot be stated this simply.

Design Patents

The patent laws provide special patents directed to new, original, and ornamental designs for articles of manufacture. These patents are referred to as design patents. Design patents protect only the appearance or shape of an article and not its structure or utilitarian features, as is the case with conventional utility patents.

Let me give an example. Suppose you've come up with a new dispenser for adhesive tape. The dispenser is shaped like a highly stylized cat, which is poised and ready to pounce. The tape comes out of its mouth. Inside the cat is a mechanical arrangement for metering the tape. The appearance of the dispenser may qualify for design patent protection. The metering mechanism may qualify for utility patent protection because of its utilitarian nature.

The proceedings relating to the granting of design patents are for the most part the same as for other patents. An application is filed with the Patent Office directed to the design. The specification of a design application is short and follows a conventional set form. Only one claim is permitted, again following a set form. The drawing of a design patent is subject to the same rules as other patent application drawings, except that no reference numerals or characters are required.

Design patent applications are subject to examination by the Patent Office in the same way as are other patent applications. If the Patent Office, on examination, determines that an applicant is entitled to a design patent, a notice of allowance will be sent to your attorney, calling for the payment of an issue fee. A design patent lasts 14 years from date of issuance.

Plant Patents

Special patents, called plant patents, are also granted by the U.S. Patent and Trademark Office. These patents are granted to anyone who has invented or discovered and asexually reproduced any distinct and new variety of plant, including cultivated sports, mutants, hybrids, and newly found seedlings other than tuber propagated plants or plants found in an uncultivated state.

An application for a plant patent consists of the same parts as other applications. The specification should include a complete detailed description of the plant and its distinguishing characteristics. The specification should also include the origin or parentage of the plant variety sought to be patented and point out specifically where and in what manner the variety of plant has been asexually reproduced. If color is a distinctive feature of the plant, it should be positively identified in the specification by referring to a designated color as given by a recognized color dictionary.

Plant patent drawings are not mechanical drawings. Rather, they are artistic drawings disclosing all of the distinctive characteristics of the plant capable of visual representation. If color is a distinguishing characteristic of the new variety, the drawing must be in color. Two duplicate copies of color drawings must be submitted. The law allows the use of photographs rather than drawings.

Since a plant patent is granted on the entire plant, only one claim is necessary and only one is permitted by the Patent Office. The oath or declaration of a plant patent, in addition to the statements required for other applications, must include the statement that the applicant has asexually reproduced the new plant variety. All papers submitted in connection with a plant patent must be submitted in duplicate. The original signed papers are retained by the Patent Office, and the copy is used by the Patent Office in requesting an advisory report on the plant variety from the U.S. Department of Agriculture.

Plant patents are in effect for 17 years from date of issuance. Additional information concerning plant patents may be obtained by writing the Commissioner of Patents and Trademarks, Washington, D.C. 20231.

And all you sexually reproducible plant fans out there, have no fear. You haven't been forgotten. The

U.S. Department of Agriculture issues plant variety protection certificates, very similar to plant patents, covering sexually reproducible plants. They last for 18 years.

7

What's Mine Is Yours—Or Is It?

Assignments, Licenses, and Infringements

WHAT'S MINE IS YOURS— OR IS IT?
ASSIGNMENTS, LICENSES, AND INFRINGEMENTS

Patent Assignments

A patent is a form of personal property, and you, as an inventor, can grant others certain rights under your patent. Let me tell you a little bit about what these things are. You should at least be familiar with them.

With a written document or instrument called an assignment, you can transfer or sell your patent to someone else. An application for a patent may also be the subject of an assignment. The party doing the assigning is called the assignor. The party receiving ownership rights in the patent or application is called the assignee. The assignee, at the moment the assignment is made, becomes the owner of the patent or application and has the same rights relative to it that the original patentee or applicant had.

With means of an assignment, the assignor can assign all or only a part interest in a patent or application. The assignment may provide, for example, for assignment of half interest, quarter interest, or whatever. Assignment interests in patents and applications are quite commonly granted by inventors to parties who have provided financial or other backing. Also, an inventor may be required, as a condition of employment, to assign rights to inventions made during the course of employment.

Since patents and applications are forms of personal property, they may also be mortgaged. A mortgage passes ownership of the patent property to the mortgagee or lender until the mortgage has been satisfied and a retransfer from the mortgagee back to the mortgagor, the borrower, is made.

An assignment or similar grant of any patent or application should be acknowledged before a notary public or other officer authorized to administer oaths or conduct other acts of a notary. The certificate of acknowledgment is prima facie evidence of the execution of the document.

Your patent attorney works with assignments on a regular basis. If an assignment is appropriate in your particular case, he or she will prepare one. Most assignments are straightforward and nothing more than a standard form. Unusual circumstances, though, could call for preparation of a more customized document.

RECORDING OF ASSIGNMENTS

After an assignment has been executed, the paper should be sent promptly to the Patent Office with a request that it be recorded. The Patent Office maintains a special department for handling document recordings.

An assignee can lose rights in an assigned patent or application if he or she does not record the assignment in the Patent Office within three months from the assignment date. If recording does not take place within the three-month period, the assignment is considered void against someone who later purchases rights in the patent or application, pays a valuable consideration for the purchase, and does not have notice of the previous assignment. In other words, the rights of the subsequent purchaser for a valuable consideration without notice are superior to those of a previous assignee. The prior assignee may, in fact, end up with no rights at all in the patent or application even though he or she may have paid the patentee or applicant for those rights. But if an assignment is recorded in the Patent Office within three months from its date, a subsequent purchaser is presumed to have notice of the prior assignment and the prior assignee who has recorded the assignment has superior rights.

An assignment, or for that matter any other document relating to a patent sent to the Patent Office, should identify the patent by its number and date of issuance as well as by the name of the inventor and title of the invention as stated in the patent. If the assignment relates to an application, rather than a patent, the application should be identified by its serial number and date of filing, and also by the name of the inventor and the title of the invention as stated in the application.

It often happens that an assignment of an application is executed before the application is filed in the Patent Office. Since the application in that situation would not yet have a serial number or a filing date, the assignment should identify the application in some other manner, as by its date of execution and name of the inventor and title of the invention. The main thing to remember is that the identification of the application should be clear enough to ensure that there is no mistake as to the application intended to be assigned.

If an application has been assigned and the assignment is recorded on or before the date the patent issue fee is paid, the patent will be issued to the assignee as owner, and the assignee of record as of that time may be noted on the patent. If the assignment is for part interest only, the patent will be issued to the inventor and assignee as joint owners.

Partial rights in a patent or application can be assigned to more than one assignee as long as the rights of the assignees do not conflict. For example, you, the patentee, may retain 50 percent of the ownership of the patent with 25 percent each being assigned to two different assignees. Obviously, you cannot assign percentages of ownership that exceed 100 percent. If you've seen Mel Brooks's movie *The Producers*, you know what kind of problems you can run into when you do that sort of thing.

A Word about Joint Ownership

As I've said before, a patent granted to two or more joint inventors will be owned jointly by them. This is also true in the case of the assignment of a part interest in a patent. Remember that any joint owner of a patent, no matter how small his part interest, can, in the absence of an agreement to the contrary, make, use, and sell the invention without regard to the other owner. In the absence of an agreement to the contrary, there can also be a sale or license of a joint owner's interest, or any part of it, without regard to any other joint owner. For these reasons, it is highly dangerous for any inventor to assign a part interest in an invention without a definite agreement between the parties as to the extent of their respective rights and obligations to each other. The advice of a patent attorney should definitely be obtained before you enter into any arrangement of this nature. Avoid grief for yourself.

PATENT LICENSES

Since the owner of a valid patent has the legal right to prevent others from making, using, or selling the patented invention, his permission is necessary before anyone else is free to do any of these things. This permission is granted by a license.

In a license arrangement, the patent owner reserves ownership of the patent to himself, and merely permits another party or parties to make, use, or sell the invention subject to conditions agreed upon by the licensor (the patent owner) and the licensee (the party to whom the permission has been granted).

A license is a contract, and it is usually, but not always, in writing. No particular form of license is required, and it may include whatever legal provisions the parties agree upon. Everything is a matter of negotiation. I've been in on a lot of license negotiations over the years. Let me mention some of the major things that have to be hammered out between the parties in almost every license situation.

COMMON LICENSE PROVISIONS

Exclusive or Nonexclusive. That is the question, or at least one of them. Do you want to give someone exclusive rights under your invention, or do you want to license several parties? In an exclusive arrangement, you put all your eggs in one basket—you are trusting in the abilities of a single licensee to reap rewards for you. With more than one licensee, you have several potential sources of monetary return. Do you want to reserve some rights to yourself to employ the invention?

Territory. Should the license grant be for the whole country, even international in scope, or should it be more restricted—for example, to a specific section of the country?

Term. How long should the license be effective? A license based on a patent cannot extend any longer than the life of that patent. But do you want it to run that long?

Nature of Use. Should the licensee(s) be able to use the invention in every way possible, or should there be some restriction as to the nature of use of the invention? For example, an invention might have several applications—be usable across a whole spectrum of products. Should the licensee(s) be restricted to only part of that spectrum?

Payment. I bet I'm grabbing your attention with this topic—unquestionably an important one to every prospective licensor I've ever met. The commonest approach is to use a royalty schedule as the primary basis for payment by a licensee. But there are other ways to go, too—like lump-sum payments, either all at once or spread out over a period of time.

Let's assume a royalty arrangement. How much can you expect? I can't answer that. What are the circumstances? I've dealt with license-agreement royalties ranging from 1/2 percent of net sales all the way to 22 percent. How valuable is the invention to you and how valuable is it to the party you're licensing it to? It's kind of like the situation involving my elderly patent lawyer friend mentioned at the beginning of the book. I have to borrow the answer he used for so many years before coming up with the photo of the concrete truck—it depends.

Regardless of the royalty rate, there are a couple of other provisions concerning payment which should be incorporated in an agreement. You should get some money up front—a licensing fee, if you will, which may or may not be applicable to running royalties. You also probably will want to have some sort of minimum payment obligations to keep the license in effect. In other words, the licensee should have to pay a certain minimum amount on a periodic basis, say, annually, to keep the license agreement alive. Without this, you could really get burned if the licensee is a dud. A similar kind of arrangement is commonly employed in exclusive agreements to keep that exclusivity. You could get burned even worse here.

Miscellaneous. License agreements incorporate many, many other types of provisions. Some fall within the "boiler-plate" category, a term lawyers use to identify agreement terms which are more or less standard. Some don't.

WHERE YOUR ATTORNEY FITS IN

Preparation of license agreements and other documents conveying patent rights should be left in the hands of a patent attorney who is a member of the bar of his state of practice. A few states have established certain formalities to be observed in connection with the sale of patent rights. In addition, federal laws exist that must be adhered to in a license agreement for it to be valid and legally enforceable.

Some of these laws pertain to the field of antitrust, and failure of parties to a license agreement to retain competent legal counsel aware of these legal restraints can result in the imposition of severe civil and even criminal penalties. Patents can also be held invalid because these laws are broken. There are also tax aspects to be considered.

No, this area is definitely not one for the do-it-yourselfer. Not only must laws be closely adhered to, but legal counsel knowledgeable in the field of patent law can incorporate specific provisions in legal documents that will maximize your rights. An experienced patent practitioner will also be aware of the various alternative approaches available to accomplish your objectives. In fact, he or she can help to define and refine those objectives. Sometimes coming up with an invention is the easy part. What do you *do* with it to maximize your rewards? That can be a tough question, and your patent attorney can help you answer it.

Let me give you a tip about where you might be able to save some lawyer costs. Bringing an attorney into negotiation meetings can be a very expensive proposition—and it's not always necessary. Often, much wasted time is involved in the negotiating process. Why have your lawyer's billing clock running while this goes on? Use your attorney as needed. Even use his absence from negotiations as a plus by deferring decision making by saying that you have to check things out with your lawyer. Please—sign nothing, agree to nothing before you do actually check with your lawyer. And, by all means, get your attorney into the process when an actual agreement document is being prepared. If he doesn't do the actual preparation, make sure that approval is given by your attorney.

INFRINGEMENT

As you've seen, the owner of a patent has the right to exclude others from making, using, or selling the invention covered by the patent. Anyone who does any one of these three things without permission of the patent owner during the term of the patent "infringes" the patent, and the owner of the patent can bring a lawsuit against the infringer. The patent owner may ask the court to stop the infringer from continuing the infringing acts. A court order doing this is called an injunction.

In addition to requesting an injunction, a patent owner may ask the court to award money damages because of the infringement.

In order to determine whether a patent is being infringed, it is necessary to determine whether or not the infringing device, process, or what have you falls within the scope of at least one claim of the patent. The patent owner has the burden of showing that the patent claims in fact do this, and an accurate comparison of the claims against the alleged infringement must be made.

The defendant in an infringement suit will, of course, try to prove otherwise. In addition, the defendant will probably allege that the patent is invalid for one reason or another and try to prove it. Although patents issued by the U.S. Patent and Trademark Office are presumed under the law to be valid, it is an unfortunate fact of life that patents are invalidated by the courts on a variety of grounds, such as failure of the Patent Office to consider the most pertinent prior art. The good news is that it is harder to invalidate a patent these days than it used to be. Patents are stronger, thanks in part to a revamping of the federal court system that handles patent matters.

If you spot what you believe to be an infringement of your patent, under no circumstances threaten or even contact the person or company that you feel to be infringing. This can have very unfortunate legal consequences. Instead, you should bring the matter to the attention of your patent attorney so that he can launch an investigation, find out whether there is in fact an infringement, and advise you as to the course of action to be followed.

It is usually faster and far, far less expensive for a patent owner and an accused infringer to resolve matters between themselves rather than to fight things out in the court system. Determining strategy at an early date is very important. A patent owner should not bring charges of infringement or threaten suit for infringement without knowing all of the facts since the party charged with infringement may itself initiate a suit in a federal court to get a judgment on the matter. It is always possible that the infringer may be able to provide some evidence tending to limit or even defeat the rights of a patent owner, so it is advisable to explore these matters thoroughly with your lawyer before actually bringing legal action.

Patent Marking

You have no doubt seen many items with patent numbers or the words "patent pending" or similar notices applied to them. What is the significance of these notices?

First of all, although it may be advantageous to do so, it is not absolutely necessary to apply a patent number to something that is covered by a patent. However, failure to mark an invention with an adequate patent notice will prevent a patent owner from recovering damages from an infringer, unless the infringer was actually notified of the infringement and continued to infringe after being given the notice. In this latter situation, the patent owner will be entitled to recover only those damages actually accrued after actual notice of infringement was given. On the other hand, a party who has infringed will be liable for damages even if it had no actual notice that it was infringing the patent if it has been making available goods identical to those carrying a patent notice.

If an invention is not yet the subject of an actually issued patent, applying the word "patented," or any other similar notice representing that a patent has been obtained, is against the law and the person affixing such a notice is subject to a penalty. This is true even when an application has been filed if the application has not in fact been issued as a patent.

The terms "patent applied for," "patent pending," or similar notices have no legal effect, but they do put others on notice that an application for patent has been filed in the Patent Office. Anyone can, however, manufacture, use, or sell an invention prior to actual issuance of a patent without being liable for infringement. These notices are of considerable value, though, as deterrents. Someone seeing them knows that a patent could issue at any time, but they don't know when. Nor do they know the scope of any patent that might issue. Producing something that could possibly infringe a patent at any time is a real financial risk.

The patent statutes prohibit use of phrases such as "patent applied for" or "patent pending" unless an application has in fact been filed and is pending. If an application goes abandoned without a patent issuing, this notice should be removed from subsequently manufactured items.

8

If You Have an Infinite Number of Monkeys . . .
The Copyright

IF YOU HAVE AN INFINITE NUMBER OF MONKEYS...
THE COPYRIGHT

WHAT IT IS

Patents can be valuable pieces of property and collectively they have generated untold billions of dollars of income to people over the years. The problem, though, as you have seen, is that patents protect only one thing—inventions.

If your creation is not an invention, as that term is legally defined, maybe it is something that can be the subject of a copyright. Fortunes have been made from copyrighted items. To give one fairly recent example, the Canadian developers of the copyrighted Trivial Pursuit game parlayed a fun, part-time project into a multimillion-dollar phenomenon—and virtually overnight too. Do you have a blockbuster game idea tucked away in your mind's recesses? Think about it. Or consider the Cabbage Patch dolls, or all the products you see around you based on the Disney, Peanuts, and Garfield characters. I'd like to have the royalties developed by any one of these copyrighted whimsical creatures in just one day. I don't know of any formal study on the point, but a darned good argument can be made that the total money generated by copyrighted works has equaled or perhaps even exceeded that from patented inventions.

The genesis of the U.S. copyright is the U.S. Constitution, which encourages literary and artistic creativity by authorizing a scheme to protect the "writings" of "authors." The current copyright statute, effective January 1, 1978, is at the tail end of a series of laws dating back to the earliest days of our republic.

The copyright is exclusively in the federal domain. Under present law there is no state copyright protection, as was formerly the case, although states still can deal with related matters such as breaches of contract, acts of unfair competition, and so on, things you'll learn more about in a subsequent chapter. Also, since extemporaneous speeches or unrecorded live performances are not protected by the U.S. government, these things, too, may be given protection by the states.

Under the single national system now in existence, copyrights are granted to authors of "original works of authorship," including literary, dramatic, musical, artistic, and certain other intellectual works. Protection is available for things both published and unpublished. Generally speaking, the owner of the copyright is given the exclusive right:

- To *reproduce* the copyrighted work
- To *prepare derivative works* based on the copyrighted work—for example, a movie based on a novel
- To sell or otherwise *distribute copies or phonorecords* of the copyrighted work to the public
- To *display* certain copyrighted works publicly
- To *perform* certain copyrighted works publicly

Subject to certain exceptions, someone doing any of these things who is not authorized by the copyright owner to do so commits an illegal act. I'll get into some of the more important exceptions a little later on.

In a broad sense you can think of a copyright as the right to copy something. It is, of course, possible that two identical "somethings" can be independently created. For example, it is said that an infinite number of typewriter-equipped monkeys having an infinite amount of time will, eventually, produce the complete works of William Shakespeare. I prefer to think that they're more likely to reproduce a Jackie Collins novel—say, *Hollywood Husbands*. Well, Ms. Collins could not successfully sue the monkeys for violating her copyright. The monkeys are not infringers. They did not copy. They could in fact publish a simian edition of *Hollywood Husbands* if they wanted to. Talk about your hairy-chested hero.

But I digress. What kinds of things are, and aren't, covered by the copyright law?

COPYRIGHTABLE WORKS

Copyrightable works include literary works; musical works, including any accompanying words; dramatic works, including any accompanying music; pantomimes and choreographic works; pictorial, graphic, and sculptural works; motion pictures and other audiovisual works; and sound recordings.

These categories include lots of things—things that you might not expect, and they should be very broadly construed. By way of example, computer programs are registrable as "literary works." So are computer data bases and compilations of data. Maps and architectural blueprints are registrable as "pictorial, graphic, and sculptural works." Well, you must be getting the idea.

I recently read a newspaper article featuring new products displaying a lot of imagination on the part of their creators. All are on the market and apparently successfully so. The interesting thing is that at least four of the five products were the kinds of things a copyright is designed to protect.

One was the Hugster, an item that looks like a bear but is described as an "indoor sleeping bag, a bed with pillow, a place to curl up, or just use as an escape." Another was the 1-Two watch which features symmetrical curves and shapes in its design. The other two were games which are successful, although they haven't reached the rarified Trivial Pursuit atmosphere as yet.

The point is that there are a lot of creative developments out there making money and they aren't necessarily patented.

A work has to be fixed in a tangible form of expression before it is given copyright protection. Something written or drawn is of course a tangible form of expression, but so, under the law, is something that is communicated with the aid of a machine or device of some kind. For example, certain types of audiovisual works fixed in machine-readable copies, such as semiconductor chips or computer-readable tapes and disks, are registrable.

Examples of things not fixed in a tangible form of expression and therefore not eligible for copyright protection are unnotated and unrecorded choreographic works, and improvisational speeches or performances not written or recorded.

The Copyright Office cannot register names, titles or brief combinations of words. A work, to be protectable by copyright, must incorporate a certain minimum of original literary, musical, or graphic expression, and the Office is of the view that names, titles, and other short phrases simply don't meet that requirement. For example, the Copyright Office refuses to register names of products, services, businesses, organizations, or groups; titles of works; and catchwords or phrases, mottoes, slogans, or short advertising expressions.

Mere ideas, methods, or systems cannot be copyrighted, although you can get a copyright on a description, explanation, or illustration of an idea or system. Let me give you an example. Suppose one day you write a book about a dog-training system that you've developed. The book is covered by a copyright and that means no one else can publish your book, its text, and its illustrations. Anyone can, however, use the idea that you've described. That's in the public domain, up for grabs, because the law doesn't give exclusivity to ideas.

Rounding out the extensive laundry list of things the copyright law doesn't protect, there is no protection for computing and measuring devices or for blank forms and similar simple items designed merely to record rather than to convey information. These things are said to lack creative authorship, which is the basic requirement of copyright protection, not only because of their simplicity but because they often consist of, or are based on, public information that is common property.

OBTAINING A COPYRIGHT

Let's assume you've created something that can be copyrighted. How do you go about obtaining copyright protection? Since January 1, 1978, it couldn't be easier. If you've created your work since that date, that is, actually fixed it in a tangible form of expression (either as a copy or a phonorecord), you probably already have a copyright, whether you know it or not.

Before 1978, statutory copyright was generally secured by publishing the work with a notice of copyright. With regard to unpublished works, you could do one of two things—either obtain a statutory copyright by filing an application with the Copyright Office or rely on a common law (state law) copyright. Starting January 1, 1978, once the first copy or phonorecord is created, federal law applies and a statutory copyright is secured *automatically* when the work is "fixed" in a copy or phonorecord for the first time.

In case you're wondering what a "copy" is under the law, it's a material object from which a work can be read or visually perceived either directly or with the aid of a machine or device. Examples are actual objects such as sculptures, paintings, books, manuscripts, sheet music, film, videotape, and microfilm. A "phonorecord" is a material object with fixed sound, such as audiotapes and disks.

PUBLICATION AND THE COPYRIGHT NOTICE

Publication of a work still has a number of important legal consequences even though current law doesn't require publication to obtain a federal copyright. The Copyright Act defines publication as "the distribution of copies or phonorecords of a work to the public by sale or other transfer of ownership, or by rental, lease, or lending." An offering to "distribute copies or phonorecords to a group of persons for purposes of further distribution, public performance, or public display" is also considered a publication.

When a work is published, each and every publicly distributed copy or phonorecord should bear a notice of copyright, a notice you've probably seen millions of times without really paying very much attention to it.

The notice for copies which can be perceived visually should have the following three elements:

1. The symbol © or the word "Copyright" or its abbreviation "Copr."
2. The year of first publication of the work (which may be omitted where a pictorial, graphic, or sculptural work is reproduced in or on greeting cards, postcards, stationery, jewelry, dolls, toys, or any useful article)
3. The name of the owner of copyright in the work (or a recognized abbreviation or generally known alternative description of the owner)

Examples are:

© 1987 John Smith
Copyright 1987 John Smith
Copr. 1987 John Smith
Copyright © 1987 John Smith

I personally prefer use of the © symbol, either alone or in conjunction with the word "Copyright" or "Copr." since the © notice gives the copyright owner certain rights in other countries under an international treaty arrangement.

A somewhat different form of notice is used for phonorecords of sound recordings. A phonorecord should have a notice with all three of the following elements:

1. The symbol ℗
2. The year of first publication of the sound recording
3. The name of the copyright owner (or a recognized abbreviation or generally known alternative designation of the owner)

Example:

℗ 1987 John Smith Music Co.

The notice is to be placed on copies and on phonorecords (or on the phonorecord label or container) in a manner and location which gives reasonable notice of the claim of copyright. The three elements of the notice should ordinarily appear together.

The copyright notice is required on published works; but what about unpublished works? As I mentioned earlier, you have federal copyright protection the instant you write down a work or "fix" it in some other way. You don't need a notice to obtain this protection in unpublished works. However, I recommend use of a copyright notice (albeit one of a somewhat different form) even for unpublished works. The form I recommend is something like this: "Unpublished Work © 1987 John Smith."

The reason I recommend the notice for even unpublished works is because of the possibility that there could be inadvertent publication without the notice. It happens all the time, especially when copies get out of the author's hands and control. Publication without the notice can destroy copyright protection. Why risk that possibility?

Speaking of publication without the notice, before 1978 even inadvertent publication by the author would destroy the copyright in his work. There was no way to salvage matters. This has been changed somewhat in the statute now in effect. Procedures are now provided to correct errors and omissions of the copyright notice. Generally speaking, the copyright will not be invalidated by omission of a proper notice if the work is registered within five years after the publication without notice and a reasonable effort is made to add the notice to all copies distributed to the public in the U.S. after the omission was discovered.

How Long Does a Copyright Last?

Good question. I'm glad I thought of it.

Under current law, for a work created on or after January 1, 1978, the copyright becomes effective for a long time—ordinarily for the period of the author's life plus fifty more years. In the case of a joint work, that is, something created by more than one person, the period runs for fifty years after the death of the last surviving author.

There are two exceptions to this general time

frame—works made for hire and anonymous or pseudonymous works (unless the author's identity is revealed in Copyright Office records). In these cases the copyright will last seventy-five years from publication or one hundred years from creation, whichever period of time is shorter. Beginning January 1, 1978, all copyrights expire on January 31 of the year of their termination. After expiration, of course, anyone is free to copy or use a work to the extent it doesn't embody a preexisting work still copyrighted.

I've just mentioned a work made for hire. What does that mean anyway? Simply that an employer or party commissioning the work is considered the legal "author" if the work was prepared by an employee within the scope of employment or specifically ordered or commissioned (and is one of nine specific types of works) and is expressly agreed in writing by the parties to be a "work made for hire." In this case the employer or the commissioning party owns the copyright outright and can apply for its registration in its name.

COPYRIGHT REGISTRATION

Although you receive copyright protection the instant you create a work, you will want to consider registering the copyright. A copyright registration is obtained by applying to the Copyright Office. The basic facts of a particular copyright then become a matter of public record. Registration is ordinarily a prerequisite to bringing a copyright infringement suit and a registration, if made before or within five years of publication, will establish prima facie evidence in court of the copyright's validity and of the facts stated in the certificate. This means that someone challenging the copyright has a much tougher time.

A registration made within three months after publication of the work or prior to an infringement will enable its owner to collect statutory damages and attorney's fees in court actions. Otherwise, only an award of actual damages and profits may be granted.

How to Obtain Your Registration

Copyrights are a do-it-yourselfer's delight. You can obtain your own registration without an attorney's help. Here's all you do. To register your work, you must send three things in the same envelope or package to the Register of Copyrights, Copyright Office, Library of Congress, Washington, D.C. 20559. The three items are:

1. A properly completed application form
2. A nonreturnable filing fee (currently $10) for each application
3. Two complete copies of the best edition of the work (one is enough in some cases)

The Copyright Office has a number of forms available, each specific to a category of works. You can obtain them either by writing to Section LM-455, Copyright Office, Library of Congress, Washington, D.C. 20559, or by calling a special Hotline that the Copyright Office maintains for this purpose. The Hotline telephone number is (202)287-9100. You can call any time to order free application forms and circulars. Orders are recorded automatically and filled as quickly as possible.

The application forms for original registrations are:

Form TX: for published and unpublished non-dramatic literary works
Form SE: for serials, works issued or intended to be issued in successive parts bearing numerical or chronological designations and intended to be used indefinitely (periodicals, newspapers, magazines, etc.)
Form PA: for published and unpublished works of the performing arts (musical and dramatic works, pantomimes and choreographic works, motion pictures and other audiovisual works)
Form VA: for published and unpublished works of the visual arts (pictorial, graphic, and sculptural works)
Form SR: for published and unpublished sound recordings

These forms are simple and incorporate instructions for filling them out.

It would also be worth your while to request the following circulars which contain good basic copyright information:

Circular R1C: "Copyright Registration Procedures"
Circular R1: "Copyright Basics"
Circular R4: "Copyright Fees"

Circular R2 lists all of the other publications available from the Copyright Office—and there are a lot of these publications.

The Copyright Office does not give legal advice, but it will answer general questions relating to copyright problems. Simply write to the LM-455 address noted above.

A myriad of legal problems can arise in the copyright area—problems such as disputes over ownership, copyright infringement, publication procedures, contracts, and royalty payments. These are the types of things you will want to discuss with an attorney. You can obtain a copyright registration on your own, but I wouldn't recommend the do-it-yourself approach much beyond that.

Mandatory Deposit

The Copyright Act has a mandatory deposit requirement for works published with the notice of copyright in the U.S. The deposited works are for use by the Library of Congress. In general, there is a legal obligation to deposit in the Copyright Office two complete copies of the best edition of a published work within three months of its publication in the U.S. with a notice of copyright. The copies in connection with the copyright registration application may be used for this purpose if the application was filed within the three-month period.

Failure to make the mandatory deposit does not invalidate a copyright. However, failure to send the deposit within three months after being requested in writing by the Register of Copyrights to do so can result in a fine.

Whole categories of works have been exempted from the mandatory deposit requirement by the Copyright Office. Obligations have been reduced for other categories. Your attorney can tell you whether a deposit is necessary in a particular case.

INFRINGEMENT

Subject to certain exceptions, any unauthorized use or copying of a copyrighted work is an infringement. Sometimes it is difficult to prove copying, but the copyright owner can avoid this by showing that the accused infringer had access to the work and that a substantial similarity exists. This usually shifts the burden to the alleged infringer to show that he didn't copy—a tough task in most instances.

Federal law, as it now exists, has carved out many exceptions to the general rule that any unauthorized use or copying of a copyrighted work is an infringement. As you've seen, use of an idea expressed in a copyrighted work is not infringement. There is also the infinite-number-of-monkeys exception—that is, independent creation of a work is not an infringement.

FAIR USE

Another important exception is the doctrine of "fair use," which allows a certain amount of copying and use for such things as comment, criticism, news reporting, teaching, research, and so forth. Factors to be considered when determining whether or not there is a fair use include:

1. The purpose and character of the use, including whether the use is commercial or for nonprofit educational purposes

2. The nature of the copyrighted work
3. The amount and substantiality of the portion used in relation to the copyrighted work as a whole
4. The effect of the use upon the potential market for, or value of, the copyrighted work

Current law also specifies circumstances under which the making or distributing of single copies by libraries and archives for noncommercial purposes does not constitute a copyright infringement.

There are a number of other specialized exemptions, including some requiring compulsory licenses from the copyright owner, that I won't go into. Your attorney will be happy to familiarize you with them. Alternatively, you can write directly to the Copyright Office for further information.

COPYRIGHT INFRINGEMENT REMEDIES

The holder of any exclusive right in a copyrighted work is entitled to bring a civil suit against an infringer. The plaintiff can seek a number of remedies. He can, for example, seek an injunction against future infringement and the impounding, destruction, or other suitable disposition of all infringing reproductions and articles used to make them. A plaintiff can also attempt to recover actual damages suffered and any additional profits of the infringer.

At the plaintiff's election, damages set out in the copyright statute can be obtained instead. Statutory damages for infringements of any one work range from $250 to $10,000, these being subject to reduction in some circumstances or to increase to $50,000 for willful infringement. A plaintiff might also be entitled to full costs including reasonable attorney's fees.

Some types of copyright infringement are criminal offenses. Willful copyright infringement for profit involving phonorecords, motion pictures, or other audiovisual works can bring both hefty fines and imprisonment to the bad guy. In addition, criminally infringing reproductions and equipment used in their manufacture can be seized and destroyed.

TRANSFER OF RIGHTS

You can license or assign rights to a copyright and, since a copyright is divisable, the party receiving an exclusive grant or license of rights becomes the owner of the copyright for those rights. Each and every transfer of rights must be in writing and signed by the owner of the right conveyed. I recommend prompt recording of instruments transferring rights in the Copyright Office since this gives constructive notice of transfer to others who might be transferred the same rights at a later time. In any event, a transfer must be recorded before the party receiving it can bring suit against an infringer.

Copyrights are powerful tools for protecting innovations, but, like patents, the innovations have to be of a certain type. If your idea cannot be shoehorned into boxes labeled "patent" or "copyright," perhaps there are other forms of protection available to you. These are discussed in the next chapter.

Before we get there, though, I'd better briefly discuss a rather specialized form of protection relating to the semiconductor-chip industry. It's not exactly a copyright, but it certainly is something close—a kissing cousin at least—and it is also administered by the Copyright Office. If you're interested in semiconductor chips, read on; if not, forge ahead to Chapter 9.

PROTECTION OF SEMICONDUCTOR CHIPS

On November 6, 1984, the Semiconductor Chip Protection Act of 1984 came into existence, which provides for the protection of mask works employed in semiconductor-chip manufacture. The owner of a mask work original—when considered as a whole, not commonplace, and commercially exploited for the first time after July 1, 1983—is given the exclusive right to reproduce the mask work and to import and distribute chips embodying the mask work. The required commercial exploitation must be in the U.S. if the owner is foreign.

The Copyright Office is authorized by law to register mask works qualifying for this protection. The owner's protection begins upon registry or upon first commercial exploitation, whichever is earlier. Protection extends for ten years if the registration takes place within two years from first commercial exploitation.

The Act provides for a special mask work notice which, in the case of a mask work fixed in a semiconductor-chip product, should be (1) on a label securely affixed or imprinted upon the package or other container used as a permanent receptacle for the product or (2) imprinted or otherwise affixed in or on the top or other visible layer of the product. A proper notice includes the words "mask work" or the symbol "*M*" or Ⓜ and the name of the owner (or a recognizable or generally known abbreviation).

The law specifically allows reverse engineering, that is, reproduction of a mask work for purposes of analysis or evaluation and incorporation of the results in an original mask work. Certain protections also are provided for innocent purchasers of infringing semiconductor-chip products.

Civil remedies under the Semiconductor Chip Protection Act generally correspond to those for infringement under the Copyright Act, but maximum statutory damages are $250,000.

9

OTHER BOW STRINGS

ADDITIONAL TYPES OF LEGAL PROTECTION FOR IDEAS

OTHER BOW STRINGS
ADDITIONAL TYPES OF LEGAL PROTECTION FOR IDEAS

SOME BACKGROUND

Generally speaking, in my opinion, if an invention is involved, the owner's best interests will usually be served by obtaining patent protection. As you have seen, though, not everything is patentable. If an idea does not meet the strict requirements of the patent statutes, what about other forms of protection? There is the copyright, of course, but are there others? Most assuredly, yes.

Courts provide protection outside the patent and copyright laws for both patentable and unpatentable inventions and ideas. This protection may be granted to the originator of the idea or invention under a variety of legal theories if certain conditions are met.

I emphasize the word *may* for two reasons. First, a good deal of the law in this area is "judge-made" or common law, that is, law based on court decisions. The fact of the matter is that judge-made law often lacks the preciseness and permanence of decision making based on statutory language like that of the patent and copyright laws. Judge-made law has a tendency to "evolve" more.

Second, the patent and copyright laws are federal laws, for the most part relatively uniformly treated by federal courts. Once you step out of the areas of patent and copyright, each state in the United States has its own jurisprudence consisting of both statutory and common law. The courts of the various states tend to handle these matters a little differently, with respect to both the theories applied to permit recovery by an originator of an idea or invention and the degree of protection that is afforded. In other words, the rights of a person who chooses not to obtain patent or copyright protection, or who cannot, are probably not as clearly defined as those of the owner of a patent or copyright.

The principal non-patent and non-copyright theories for recovery used by the courts to grant relief to the originator of an idea or invention are set forth below. Whether one or all of these theories are alive and well in your particular locality is a matter that should be taken up with your patent attorney along with the question of their applicability to the facts involved in your particular case.

THE CONFIDENTIAL RELATIONSHIP

If a party, such as a company, receives information concerning an idea in confidence, courts will usually provide legal relief to the submitting party if the information received is not maintained in secrecy or if the receiving party adopts and makes use of the idea. Although an idea need not be patentable to be afforded this type of protection, it is usually required that it be new or novel. Most courts are of the view that there can be no property right in an idea already known to the party receiving the disclosure or in a matter of general public knowledge. Also, if the idea is easily obtainable through inspection of a product already being sold or if it is of an obvious nature, there is usually no recovery based on its use.

A promise to keep an idea secret or confidential need not necessarily be in writing to be enforceable. In fact, a court may find a promise implied merely by the acts or conduct of the parties involved. There need be no verbal announcement whatsoever by the receiving party that the information will be treated as confidential to find a requirement of confidentiality in such cases.

Here's an example. Someone with an idea offers to disclose it to a company along with the announcement that the disclosure is considered to be confidential. If the company indicates that it would like to receive the information, saying nothing about confidentiality, a court is likely to infer from the company's conduct alone that it was indeed accepting the information under the terms established by the submitter; that is, in confidence. The court would likely do this in the interest of equity to prevent unjust enrichment of the company were it to adopt the idea and use it for its own benefit.

The relationship between the parties will occasionally lead courts to find an obligation on the part of the recipient of an idea to keep it in confidence. If the two parties are dealing on unequal terms which result in the submitter reposing trust and confidence in the other's good faith, courts will tend to find a confidential relationship if that good faith is abused. What I'm talking about here in many of these cases is the big guy apparently taking advantage of the little guy—an element of unfairness. Judges and juries are only human and you'd be foolish indeed to think that these cases get decided on legalities alone.

All of these cases, of course, depend upon the specific facts involved, and each must be determined on its own merits. A solid understanding of the current state of the law in the particular jurisdiction involved is absolutely essential for one to fully assess a submitter's rights in any given situation. You should consult legal counsel before yelling "Stop, thief!" based on a belief that your idea has been misappropriated. Even if a confidential relationship is found to exist, courts will be unlikely to find liability on the part of the recipient unless the idea was in fairly concrete form and actually used by the recipient or disclosed by it to another party in breach of the confidential relationship.

Express Contract

Parties can enter into an express oral or written agreement whereby one of the parties agrees to disclose an unpatented invention or other idea to the other in return for monetary or other compensation if the receiving party ends up using the idea. The compensation may be specified in a dollar amount, or it may be defined in more general terms such as "reasonable compensation."

Many courts will enforce contracts of this nature, if the idea is of sufficient value to support a contract. Generally speaking, an express contract will not be enforceable if the idea is overly vague or abstract or if it is not new.

There is an interesting case on this latter point involving a well-known cleanser company. The company entered into an agreement with a man claiming to have a surefire way to increase profits. The company agreed to pay the man a sum of money to hear his suggestion. The suggestion, believe it or not, was that the company raise its prices. With the passage of time, the company did, of course, raise its prices. After the prices were raised, the man tried to collect the sum he felt was owed to him. The company refused, and the man went to court to enforce his contract. The court said, in effect: "Sorry, Charley, but the raising of prices ranks right up there with death and taxes insofar as the inevitables of life are concerned." The idea was felt to lack novelty, that is, it wasn't new, and the contract was held not to be enforceable.

Although a contract itself may not state it in so many words, courts are reluctant to allow recovery by a submitter of an unpatented invention or other idea if the recipient does not make actual use of it. Also, if the submitter has disclosed the idea before the recipient has promised to pay for it, a contract entered into after such disclosure may not be enforceable because the disclosure has already been made and won't constitute a valid consideration to support the contract.

IMPLIED CONTRACT

An implied contract differs from an express contract in that the consent of the parties to the contract is indicated by their actions or conduct rather than by their words. Ideas may be submitted under circumstances which imply a promise to pay by the party receiving them. Suppose, for example, a company actively solicits new ideas by running some newspaper ads. In response, Bernard Brainstorm discloses a new idea that is in fact later put into use. Courts are prone in solicitation cases like this to find an implied promise to pay. If a designated amount of compensation has not been mentioned, the courts will probably find an implied promise to pay a "reasonable" amount of compensation.

A variation of this implied promise to pay is the situation in which the submitter approaches the prospective recipient and says to him: "I have an idea. I am willing to submit it to you, but if you use it I expect to be paid a reasonable amount for it." The recipient makes no actual promise to pay, but merely says: "Let me see your idea."

The submitter in such a case will usually be found entitled by the courts to receive reasonable compensation if the idea is in fact disclosed and used. Note the distinction here from the "blurted-out" situation where the submitter is entitled to no compensation if he or she discloses the idea without being asked to and without giving advance notice to the recipient that compensation is expected.

There are many other fact situations where the implied contract doctrine may come into play to protect an unpatented invention or other idea. Such factors as the relationship between the parties, the business or industry involved, and many others will be taken into account by a court before it reaches a determination that an implied contract exists. As with express contracts, ideas that are not new or not concrete may fail to support an implied contract. Also, as you might expect, implied contracts are more difficult to prove than express contracts, especially written express contracts.

Trade Secrets

Maybe your idea qualifies for protection as a trade secret. Again, the question of whether or not this legal theory has application depends upon the nature of your idea. Trade secret coverage is usually applied to things like formulas, customer lists, methods of doing business, manufacturing processes, and technical know-how: the kinds of things that provide some commercial advantage and aren't known generally to the public.

The best-known product in the world, Coca-Cola, is a trade secret, so it shouldn't be news that trade secrets can be valuable property rights. Someone misappropriating a trade secret is subject to a civil lawsuit for damages and, in some cases, even criminal prosecution.

Let me give you a couple of trade secret pluses first. A trade secret can last forever, as long as it remains precisely what the name says—a secret. Compare that with the limited life of a patent. The Coca-Cola Company might not even exist today if the formula for its main product had been patented by its inventor instead of being locked up in a vault. After the patent expired, competitors, lots of them, would have popped up overnight with the identical product described in the patent.

Another plus for the inventor. Trade secrets can cover a whole spectrum of things, including things that could not be patented. Patents, on the other hand, must fall within specific classes of invention defined by federal patent law.

Yet another plus in the trade secret's favor. Patents can be expensive to obtain. A trade secret can get as cheap as it's possible to get—clear down to nothing. No filing fees, no maintenance fees. Not even attorney's fees until you go to enforce the trade secret. Just the costs associated with keeping the thing you want to protect secret.

Okay. If trade secrets are so great, why waste your money getting a patent? Well, as it turns out, for several very good reasons. Now come the cons.

Here's one—a big one. Trade secrets exist only so long as they stay a secret. Many things, the vast majority actually, can be easily copied once they see the light of day. A simple mechanical device, an electrically powered nutcracker say, could be readily copied once the first one hit the streets. The so-called trade secret would have a very limited life span indeed. Not everything is as hard to reverse engineer as the Coca-Cola formula. Few things are.

Another big negative. A trade secret can only be enforced against someone who wrongly acquires or appropriates it—this could be someone, for example, who steals the secret from you, or who promises to keep it secret and doesn't. If someone merely copies your idea after it hits the marketplace or comes up with the same idea independently on his or her own, there's nothing in the world you can do about it. With a patent, you can prevent *anyone* from making, using, or selling your invention.

TRADEMARKS

A trademark is a word, a design, or a combination of word and design used to identify goods or services of a particular manufacturer or merchant and distinguish them from those of others. Words and designs are forms of ideas, sometimes potentially very lucrative ones, so you should know where trademarks fit into the general scheme of things.

Rights in a trademark are obtained by using it on or in connection with goods (products) or, in the case of services, by displaying it in connection with sale or advertising.

If a party uses another's trademark (or one confusingly similar), it might be an infringer and be subjected to penalties under both state and federal law. Whether or not there is an infringement depends upon such things as whether the mark was used on the same or closely related goods and services and whether use was in the same geographical area or natural area of expansion. Another factor is whether the mark is "strong" or "weak." A strong mark is one that is well-known and distinctive—one that has become closely identified by the public with a particular company or other source of goods or services. "Kodak" comes to mind as a very strong mark, strong enough to be enforced against anyone using "Kodak" for just about anything, even products widely removed in character from Eastman Kodak Company's product line. On the other hand, some marks, such as "Acme," are weak, being used by many different parties for many different products and services.

Both the federal government and the various individual states have statutes permitting registration of trademarks. While trademarks need not necessarily be registered to be protected, registration confers many important advantages. This is especially true for federal registrations, which are granted by the U.S. Patent and Trademark Office.

A federal registration, among other things, (1) gives constructive notice of the registrant's claim of ownership; (2) establishes nationwide protection for the registered mark; and (3) provides rebuttable presumption of the registrant's exclusive trademark rights. Your attorney can acquaint you with other advantages of federal registration.

A trademark right can last forever. The only requirement is that it continue to be used. Federal registrations, however, have a limited life span and must be renewed every twenty years. Renewals can continue indefinitely as long as the owner continues proper use of the registered mark. Under federal law, interruption of use of a trademark for two years is ordinarily considered an abandonment. It's a case of use it or lose it. Incidentally, if a mark is abandoned by an owner through non-use, rights in the mark will be acquired by the first subsequent user.

You may be puzzled by the various notices you see used in association with trademarks. A ™ means that the user claims common law (nonstatutory) trademark rights. The ℠ designation denotes common law rights for a mark used to identify a service. The ® notice means that the mark is federally registered. A notice such as "Registered U.S. Patent and Trademark Office" means the same thing. Registration notices do not have to be used, but they do provide a number of advantages that your lawyer can tell you about, particularly when it comes to collecting damages for infringement.

An owner can license someone else to use his trademark but is legally required to control the nature and use of the trademark by the licensee. Failure to exercise this control will generally result in complete destruction of the owner's trademark rights.

Assignment of a trademark is also permitted, but here, too, there is a quirk in the law. Seems like the woods are full of them. Maybe that's why we have so many lawyers. Anyway, an assignment of a trademark must include the business and good will associated with the mark.

A trademark infringement is a form of unfair competition, a relatively nebulous legal area encompassing a great deal of judge-made (common) law.

COMMENT

You're probably a little confused by now about the various kinds of things covered by trademarks, patents, and copyrights, as well as by the other forms of protection I've described. It probably seems that there is some overlap—that a particular type of idea could be covered by more than one of these legal forms of protection. The fact of the matter is, you are absolutely right. Sometimes, depending upon its nature, an idea can qualify for several forms of protection. If it has the requisite artistic merit, the idea will qualify for copyright protection. If the idea relates to an article of manufacture that is new, original, and ornamental, a design patent might be in order. If the features are functional, a conventional utility patent would be appropriate. And finally, if the idea incorporates a specific design which operates to identify the product and distinguishes it from those of others, it can function as a trademark.

Okay. Now that you have a good, thorough grasp of these concepts, I'm going to let you play patent lawyer. Imagine yourself sitting in your office one day cataloging all of your hard-won diplomas and professional certificates.

There's a knock on the door. The person walking in is a prospective client who has shown up without an appointment. Somehow, the man has managed to make it past Miss Murchison, your vigilant and highly protective secretary.

The man is carrying an odd-shaped something wrapped in brown paper and baling twine. He's willing to show the odd-shaped something to you if you promise to keep it secret. You cross your heart and hope to die if you ever open your yap without the man's permission. He's satisfied. Slowly, ever so slowly, he unties the twine. Slowly, ever so slowly, he unwraps the paper. Tension fills the air, just like at the end of *The Maltese Falcon* when that fabled bird was unpacked in Sam Spade's office.

Suddenly, there it is. The object is displayed before you. It, too, is the stuff that dreams are made of—an inventor's dreams. You study the thing for a moment, and then put to the visitor the question patent lawyers are famous for. "What is it?" you ask.

It's a salt lick, the man explains. He's a rancher in Washington State and he's had problems with conventional salt blocks. His cattle don't like the flavor the blocks have and the blocks have problems in the rugged environment, where it seems it's always raining. Conventional salt blocks erode away under the near constant downpour.

"I'd like to protect my new block," the man says. That sounds reasonable, you think. But what forms of protection can he get? You study the block. You ask questions. You determine the following.

The rancher's block has a specific new formulation that resists erosion and tastes particularly good to cattle. The formulation, you decide, may very well qualify for utility patent protection.

The rancher's block has a rather peculiar shape. It seems that the rancher is a sculptor of sorts. He wants to protect the shape of the block, which apparently has no particular function insofar as its operation is concerned. "Aha!" you declaim. The block appears to be new and ornamental. Perhaps the appearance can be covered by a design patent. Perhaps the block can also be protected under the copyright law. After all, it is "original" and incorporates some degree of artistic creativity.

There is even the possibility of trademark protection. You note that the block is molded with the man's initials going clear through the object in contrasting colors. This could certainly function to identify the manufacturer of the block when it is sold, and that is, of course, the function of a trademark.

It is unusual for a particular thing to simultaneously qualify for protection under the patent, trademark, and copyright laws, but, as you can see by the example I've just given, it's not impossible.

Sometimes, though (in most cases, actually), only one form of protection is appropriate. The thing you want to protect may be suitable only for a patent—or for a copyright—or for a trademark.

Then, too, it is not uncommon for various forms of protection to be inconsistent. Some people in our society today are determined to have it all. Well, in this particular field you can't always satisfy that desire.

Let me give you some illustrations of what I mean when I talk about inconsistent forms of protection. Rather than try to obtain a utility patent on the salt block formulation for your rancher client, you might have suggested protecting the formulation as a trade secret. Obviously, you could not do both since an issued patent is a matter of public knowledge. You have to make that disclosure as the price paid for the patent monopoly. Something known to the public can't be a trade secret.

Or what if your client had a computer program he wanted to protect. A conventional statutory copyright notice is applicable only to published works and it might very well be inconsistent to place an unexplained notice of this type if the client considered the program to be secret information. Publication and secrecy are essentially at odds with one another.

Sometimes the appearance and configuration of an article can be a proper subject for a copyright or a design patent, and can also function to identify a product and distinguish it from those of others; in other words, qualify as a trademark. One well-known object that is a registered trademark is the famous hourglass-shaped Coca-Cola bottle. And would you believe that someone could obtain a federal trademark registration on the concept of placing a cloth tab on a back pocket of a pair of pants? True. Levi Strauss has obtained this form of protection—and enforced it too.

It is not always easy to tell what form (or forms) of protection might be right for you and appropriate to a particular set of circumstances. This is a job for a real patent lawyer, not a pretend one. But now you have a good understanding of what your options might be.

10

TAKE MY IDEA ... PLEASE

SUBMITTING YOUR IDEA TO A COMPANY

TAKE MY IDEA ... PLEASE
SUBMITTING YOUR IDEA TO A COMPANY

SOME GENERAL COMMENTS

Every day companies in the United States receive thousands of ideas and suggestions from customers and other non-employees. Some of these are, of course, submitted without any expectation of payment, while others are sent with the thought that there will be money or other compensation if the company adopts and uses the suggested approach. These submissions may relate to new products, to advertising ideas, or to practically any other aspect of a company's activities.

I used to review all of the ideas submitted to a major manufacturing company, a dozen or so a day, and I can tell you from my own personal experience that only a small percentage of ideas submitted to companies are adopted by them for one reason or another. The idea may be impractical from a manufacturing or marketing standpoint, or it may in fact be nothing new—at least not new from the company's point of view.

In spite of this, most companies are receptive to new ideas and will review these ideas to determine whether or not they are of commercial interest. Just as prospectors are willing to process tons of sand to find an ounce of gold, so too are most businesses willing to look for that one idea out of many that may pan out—a nugget which will be a valuable asset to the company's operation.

Such an enlightened attitude does pay off. Xerox Corporation became Xerox Corporation, for example, because when it was a small company called Haloid it gave favorable consideration to an idea proposed to it by a man named Chester Carlson. Mr. Carlson's idea related to a way to make dry copies by applying a powder to an electrically charged plate. He called the method xerography. In the same city of Rochester, New York, another example. Eastman Kodak received the idea for one of its major film lines from two musicians, guys who did most of their photographic processing in a bathtub.

If you plan to submit an idea to a company, or to make a suggestion to a company, you should, however, be aware of a fact of life, and that is that most companies, especially the large ones, tend to treat submissions as potential lawsuits. The companies aren't being paranoid either. Over the years there have been many lawsuits filed by people who felt their ideas had been ripped off. Remember my story about the fellow who suggested that a company raise its prices in order to increase profits? As someone once said, occasionally paranoids really are threatened.

Most big companies have established a procedure for handling outside ideas, inventions, and suggestions to eliminate, or at least minimize, their liability based on exposure to submitted ideas.

A number of companies (a minority) refuse to even consider submissions not covered by either patent applications or issued patents. An even smaller minority go so far as to refuse to consider outside submissions not covered by actual issued patents. The majority of companies will, however, consider all manner of ideas, suggestions, and inventions, whether patentable, copyrightable, or not, although most, before actually considering the matter, will ask the submitter to sign a document establishing conditions under which the review will be made. This subject will be explored in greater detail a little further on in this chapter.

SELECTING A COMPANY— FACTORS TO CONSIDER

To avoid a great deal of wasted time, effort, and money, if you wish to submit an idea to a company, do your homework. Select the company or companies most likely to have an interest in the suggested approach. To this homework should be added a liberal sprinkling of common sense.

It most cases you will probably wish to eliminate from the list of companies to be contacted those that do not have a product line at least somewhat related to the idea. To use an extreme example, if you have an idea for an automobile transmission you'd only be wasting your time contacting a company whose sole product is toothpaste.

Thomas Register, which is probably available at your local library, is an extremely helpful reference in this connection. *Thomas Resister* is a multivolume publication listing a large number of products and services alphabetically, along with names of the companies that provide those products and services. A similar breakdown is provided by *Standard & Poor's*

Register of Corporations, Directors and Executives. Both of these registers also provide the addresses of the various listed companies.

Dun & Bradstreet puts out a couple of reference works you might be interested in taking a look at—*Million Dollar Directory* and *Middle Market Directory*. These publications provide information on many thousands of companies with net worths exceeding $500,000.

Another valuable source for names of companies dealing in certain product areas are trade directories and magazines. There are literally thousands of these publications geared to a specialized reading public within precisely defined areas of interest. Your local librarian can assist you in locating the correct publication for your purposes.

If your idea involves consumer products, such as household appliances or personal-care items, you will probably be able to put together a pretty good list of companies having a possible interest in the idea by keeping an eye peeled for advertisements directed to similar products and by visiting stores. A few inquiries made to businessmen, attorneys, and bankers whom you know may also generate leads to likely prospects. Check out product distributors—middlemen—in your area of interest. It's a surprisingly small world out there. People working in an industry know that industry, its trends, needs, and state of development.

While it is true that you will be more likely to find a good reception for an idea by those companies having an established related product line, it is possible that a particularly aggressive, expansion-oriented company will be willing to take on a new development if it is good enough, even though it appears at first blush to be outside its regular area of interest.

These situations are unusual, however, and you will normally find it advantageous to stick with companies having at least marginally related product lines and areas of expertise. A company of this latter type will have established markets for a suggested product and will, therefore, be more likely to sell it successfully. Also, such a company will be familiar with the problems in the area, and will be able to provide you with a realistic appraisal of the idea or invention. A company already making available a product to which your suggestion pertains will be less likely to pass up something really good, something which will sharpen the company's competitive edge in its established markets.

You should also learn something about the reputation of the company with which you are considering dealing. Companies, like people, do have reputations that have been built up over the years. An organization that has been innovative and periodically introduces new products is much more likely to be receptive to new ideas than a stick-in-the-mud company that is conservative in this regard, a company tending to stay with established product lines in essentially unmodified form. Quite frankly, a company of this latter type probably has a case of corporate hardening of the arteries and isn't worth contacting anyway. I know. I used to work for one. To me there's no surer sign of a company on a downhill slope than a lack of innovation.

Company annual reports and information sheets obtainable from stock brokerage houses often contain information which will throw some light on this point. If a company has a substantial research and development budget, it is almost a foregone conclusion that it will also be receptive to ideas pertaining to new products and developments. Another key indicator of how innovative a company is would be the number of patents that it has assigned to it. The Patent and Trademark Office maintains lists of companies and the patents assigned to each. These lists are available for public inspection.

It's important to check out the economic health of the companies with which you are considering dealing. You want to hitch your wagon to a thoroughbred champing at the bit to be first across a finish line, not a worn-out old plug likely to pull it no farther than the glue factory. The sources of information I just mentioned—the annual reports and brokerage information sheets—will help you put your finger on a company's economic pulse.

Other sources of information are boards of trade, chambers of commerce, and similar commercial organizations in the geographical area of the company under consideration.

Another factor that may be of interest to you is the location of the company. There are some very real advantages to dealing with a company near you. First, proximity between the submitter and the company makes it convenient to arrange for personal meetings with the appropriate company representatives. Personal meetings are usually much more effective from the standpoint of getting an idea across. In addition, the company is likely to give greater consideration to an idea from someone with whom it has been in direct personal contact than one from a person who has submitted it by mail. Pressing the flesh and eyeball-to-eyeball contact are likely to pay off with a more thorough review of your idea.

Geographic proximity also provides greater opportunity for an exchange of views between the submitter and the company, and questions can usually be answered far more quickly and more satisfactorily through the give-and-take of discussions rather than by means of written correspondence. A submitter's enthusiasm for the idea is much more likely to rub off on company officials during personal meetings and conversations than through the medium of written correspondence.

People dealing with companies in their own areas may have acquaintances working for the organizations in a position to open the right doors and ensure that the idea receives serious consideration. Or it may be that mutual business associates of the submitter and the company representatives exist who may be able to assist in getting the parties together. Then, too, don't discount the advantages of knowing an executive's Aunt Minnie or belonging to the same club or church.

The size of the company is yet another factor you should take into account. Bigger is not always better. On the one hand, a large company is an established instrument for manufacturing and marketing a product. On the other hand, a product suggested by someone from outside the company may merely get lost in the shuffle or become a victim of the dread "not-invented-here" syndrome, which simply means that some corporate research people turn a deaf ear to ideas they themselves did not generate.

It may be that a smaller company will be in greater need of a new product for its continued economic health and well-being than a large company, and better able to effectively direct its energies into getting your product on the market. These things, must, of course, be fed into your decision-making process with other considerations such as the economic viability of the small company.

Since you are less likely to encounter bureaucratic red tape and rigid adherence to an established procedure for handling new ideas when you deal with a small concern, you may not be asked to sign a release form of the type commonly used by most larger companies. A smaller company might even agree to hold an idea in confidence or promise payment if it uses an unpatented idea, a situation that the submitter is not likely to find when dealing with a large established concern. It is, however, difficult to provide any hard and fast rules in this area.

Remember that you are free to simultaneously submit your idea to as many companies as you wish. Some people are under the impression that they have an obligation to restrict their idea submissions to only one company at a time. This is incorrect. While you may, of course, contact only one company, in the absence of an agreement to the contrary you are by no means obligated to do so.

How to Submit a Patented Invention to a Company

Let's assume you've selected a company to contact concerning your idea. If that idea is in the form of an invention and an issued patent covers it, the matter of submitting the patent to a company for its consideration is very straightforward. As you will see, the situation becomes more complex when you have a mere idea or an unpatented invention (including one covered by a pending application).

Since a patent is a matter of public record, a company assumes no legal risk by receiving and reviewing a copy of a patent submitted by an inventor. For this reason, a company will not normally ask a person submitting a copy of a patent to sign a release agreement, as is commonly the practice with respect to ideas and unpatented inventions. The patent owner, on the other hand, relies for protection on the issued patent and usually doesn't find it necessary to establish any other prior conditions to protect his legal interests.

When approaching a company to see if it has an interest in an issued patent, it is conventional practice to send a copy of the patent under cover of a letter spelling out just what the writer has in mind.

As you've seen in Chapter 7, you can sell a patent, assign it, or grant licenses under it. If you are submitting the patent to a company with only one of these alternatives in mind, you will want to make this clear in the cover letter. If, on the other hand, you have an open mind on the subject, the simplest and best approach is merely to forward a copy of the patent to the company with a statement in your cover letter to the effect that you will consider any proposal relative to the patent that the company may wish to make to you. If you have financial terms in mind for either assignment of or licensing under the invention, you will want to put these in your cover letter.

Rather than enclosing a copy of a patent with their letter, some people just refer to the patent number in the text of the letter, leaving it to the company to obtain its own copy. This is not good practice, primarily because of the time delay involved. Also, it is simply a matter of proper business etiquette for the patent owner to furnish a copy of the patent to a company for its review. A good copy run off on a duplicating machine will do. Anyone is free to make copies of patents since the patents themselves are considered to be in the public domain insofar as duplication is concerned.

If you merely plan to bring your patent to the attention of a company, and not make specific proposals, you take no risk by approaching the company on your own to see if it has an interest in the invention. If, however, you wish to set out definite terms to be offered to a company in your cover letter, you should fully explore with your patent attorney the various alternatives available in this regard before writing. In addition, if you feel that the company is infringing your patent, you should positively, absolutely, by no means, never, ever make contact without first obtaining legal counsel. I hope I have made myself perfectly clear. Serious legal implications arise out of an accusation of infringement, and such a step should not be taken by a patent owner without considering it fully in consultation with a patent attorney. For one thing, an accusation of infringement immediately opens the door for the accused party to initiate a lawsuit challenging the patent.

An appropriate form of a letter to a company by a patent owner who has not established terms and merely wishes to bring his patent to the attention of a company is as follows:

> Mr. Smith
> X Corporation
> Address
>
> Dear Mr. Smith:
>
> I am the owner of U.S. Patent No. [fill in number], a copy of which is enclosed.
>
> I am submitting this copy to you with the thought in mind that the patented invention may be of interest to you. I am willing to consider granting your company a license under this patent or selling the patent to you under terms to be negotiated.
>
> I look forward to receiving a reply from you within the near future.
>
> Yours very truly,
>
> Jane Jones

This form letter is not cast in bronze. You may want to supplement it with more information about the invention. Be careful, though, not to say anything against your own best interests.

The letter should be specifically addressed to *someone* at X Corporation. Who? I think the president of X Corporation, that's who. Now don't expect the answer to come back from the president. It will probably come back from someone else, the company's patent counsel maybe. You're not sending the letter to the president to get his autograph. You want action, and when a company employee gets a buck slip from the office of the president, he gives whatever is attached to it top priority. You can find a company's president's name in a number of sources, including the corporate annual report, the *Standard & Poor's Registry of Corporations, Directors and Executives*, and Dun &

Bradstreet's *Reference Book of Corporate Managements*.

After receiving a letter of this type and the accompanying patent copy, receipt will normally be acknowledged by the receiving company. The patent copy will probably be circulated within the appropriate manufacturing and marketing groups within the company to see whether or not they have an interest in the patented invention. Also, if the patent is of interest, the company will probably launch an investigation, determining to its own satisfaction that the patent is valid. The company will look into the patent's scope and applicability to its operations.

After the company has considered the merits of the patent, it undoubtedly will let you know whether or not there is an interest. This could take a month, or much longer. If there is an interest, the company may quote licensing or sale terms or, alternatively, it might suggest that you spell out acceptable terms. You will recognize the fact that guidance from a patent practitioner experienced in licensing and related matters is invaluable at this stage.

How to Submit Unpatented Inventions and Ideas to a Company

While submitting a patented invention to a company is relatively straightforward and uncomplicated, the submission of unpatented inventions and ideas to companies presents special problems.

While a patented invention may be disclosed to a company without establishing prior conditions for disclosure, this is not the case with unpatented inventions and ideas. Before making a disclosure of this type to a company, you will first want to establish the conditions under which the idea or unpatented invention is to be considered.

If at all possible, you will wish to establish a confidential relationship with the company. That's what *you* want, not what the company wants. Rarely will a large company agree to hold submitted material in confidence. But let's operate under the theory of nothing ventured, nothing gained. Make your first letter to a company concerning your idea or unpatented invention along the following lines:

> President
> X Corporation
> Address
>
> Dear Ms. Smith:
>
> I have an idea [invention] that I would like to submit to your company for evaluation purposes.
>
> I consider the information relating to this idea [invention] to be secret, and in consideration for receipt of this information from me, X Corporation agrees to keep such information in confidence and not to use such information without receiving my written permission.
>
> If after receipt of the information, X Corporation determines that it is interested in the idea [invention], we will negotiate in an attempt to arrive at mutually agreeable terms for its use.
>
> If X Corporation is agreeable to receiving information concerning the idea [invention] in accordance with these terms, please so indicate by having an authorized company representative sign the enclosed copy of this letter at the place indicated below and return it to me. I will then disclose details of my idea [invention] to you.
>
> Yours very truly,
>
> James Jones

Agreed to and accepted this
____ day of _____, 19____
By: _____
Title: _____

I will not mislead you and tell you that your chances of receiving the requested signed copy of the letter are very good. Most companies, and in particular the larger companies, refuse as a matter of corporate policy to receive disclosures in complete confidence. However, all things are possible in this world, and there is at least a chance that you will find a company that is willing to go along with the terms that you have established in your letter.

More than likely, though, this will not be the case, and you can probably expect to receive a letter from the company generally informing you of the terms under which it is willing to consider ideas and unpatented inventions. The terms won't be the ones you proposed. The company's letter is likely to be accompanied by a disclosure agreement stating its terms in detail. The company will ask you to sign this agreement before it will consider your idea.

While these corporate agreement forms vary in length and complexity, they boil down pretty much to the same thing: that the company will refuse to consider the idea unless the submitter agrees that the review will be on a *no-obligation, nonconfidential basis,* with the company reserving the sole right to determine what compensation, if any, it will pay for the use of the idea.

Reprinted below are the conditions set forth on a form sent out by a major U.S. company. These terms are typical, and I am sure that after you read them over you will consider them harsh. Before passing judgment, though, you should consider the company's side of the issue. It's a fact of life that many ideas submitted by customers and others often relate to things the company knows about. Perhaps the company has been working on a related project in its own laboratories. It might even be poised for introduction of the product into the marketplace.

The company is interested in protecting itself so that it can continue with any plans it might have without being subjected to a legal action brought by a submitter under the mistaken belief that the idea has been ripped off. Of course, even in the absence of a signed release agreement a company could, in all probability, avoid liability by proving that it had the idea first; but you can understand the reluctance of companies to enter into protracted and possibly costly litigation to prove this point.

Here are one company's conditions:

CONDITIONS OF SUBMISSION

1. No confidential relationship is to be established or implied from consideration of the submitted idea or the material submitted in connection with it. The material is not to be considered as submitted "in confidence" and the Corporation makes no commitment that the idea or material shall be kept a secret.

2. If the idea submitted to the Corporation is protected by a patent and if the Corporation desires to obtain rights with regard to such patent, the terms applicable to its obtaining such rights shall be subject to negotiation between the Corporation and the patent owner. The consideration of an idea by the Corporation, however, shall not impair the Corporation's right to contest the validity of any patent that may have been or may thereafter be obtained on it.

3. In the event the idea submitted to the Corporation is not protected by a valid patent, it will be within the Corporation's sole discretion to determine the extent of any value to the Corporation of the idea and of any compensation which may be paid for its use by the Corporation.

4. The Corporation will give each submitted idea only such consideration as in the judgment of the Corporation it merits and the Corporation shall be under no obligation to return any submitted material.

5. The Corporation shall be under no obligation to reveal any information regarding its activities in either the general or specific field to which the submitted idea pertains.

6. The Conditions of Submission stated herein are controlling and any prior agreements, either express or implied, which conflict with such Conditions of Submission shall not be binding on the Corporation.

Well now, what should you do? You have the company's form in front of you, and while you can appreciate their side of the issue, you would still like to submit your idea under terms more favorable than those the company has outlined for you in its disclosure agreement.

Although some companies will under no circumstances budge from the conditions set forth in their idea submission agreements, others may be willing to modify their stance somewhat. Some companies, when pressed, will accept ideas and unpatented inventions on a confidential basis, assuming that their obligation of confidentiality is limited to a specific period of time and subject to certain exceptions.

If you receive an idea disclosure form from a company or if you receive a letter from a company stating that it will not receive ideas or unpatentable inventions under the terms of the letter that you originally sent it, you might possibly persuade the company to modify its position. Put the ball back in the company's court, but put a little spin on it. Send out another letter, this one suggesting terms *somewhere in the mid-range* between those of the company and the terms of your first letter. Address it to the person who wrote to you, but copy the president. A sample letter to X Corporation taking this tack would be as follows:

> Mr. Black
> X Corporation
> Address
>
> Dear Mr. Black:
>
> In reply to your letter of [date], I would like to propose that the submission of my idea [invention] to you be in accordance with the following conditions.
>
> X Corporation agrees to maintain in confidence and not to use in its affairs all information received from me relative to my idea [invention]. This obligation shall extend for a period of [fill in number] years from the date this letter is signed on behalf of X Corporation as noted below. In no event shall X Corporation be required to keep secret and maintain in confidence information relating to my idea [invention] which:
>
> (1) is already known to X Corporation prior to receipt thereof from me, as proven by written records of X Corporation;
> (2) is or becomes generally available to the public through no fault of X Corporation; or
> (3) is received by X Corporation from a third party under no obligation of secrecy to me.
>
> If X Corporation is agreeable to the above conditions, please so indicate by having an authorized representative sign at the place indicated below. I will then send you information concerning my idea [invention].
>
> Very truly yours,
>
> James Jones

Agreed to this ____ day
of _____, 19____
On Behalf of X Corporation
By: _____
Title: _____

cc: Ms. Smith

X Corporation may pleasantly surprise you and return a signed copy of your letter. In that case you have the protection of the terms set forth in the letter insofar as your disclosure is concerned. If X Corporation does not accept your counter-offer, you always have the option of filling out the form the company sent to you previously and making your submission on those terms.

It is most important to bear in mind, though, that submission of an idea on a nonconfidential basis will start the one-year term during which a patent application must be filed. If you're hazy on this I suggest that you reread Chapter 6. Another zinger exists, too. A nonconfidential submission of your invention before it is the subject of a filed U.S. patent application might prevent you from obtaining valid patent protection in many foreign countries. Don't do anything rash at this juncture without touching base with your patent attorney.

If you find X Corporation's terms too repugnant, you may decide to stop dealing with that company altogether and approach Y Corporation with your idea. This is your choice. You must always consider the possibility, however, that no company will accept your terms, so you may have to accept what you feel to be harsh terms in order to get your idea considered.

AFTER DISCLOSURE CONDITIONS HAVE BEEN ESTABLISHED

Let's assume that you and X Corporation are now in written agreement on the terms under which you will disclose your idea. What happens next? The obvious answer is that you will make your idea known to the appropriate people at X Corporation. How this is to be done is something that you will have to work out with your contact at X Corporation.

You may, of course, either mail your idea to the company or present it to the company in person. As I have already mentioned, I prefer personal meetings, if feasible, because I think they are more effective. If this is the way you want to go, try to arrange a meeting at the earliest possible date with those representatives of the company who are in a position to make decisions on behalf of the company relative to the idea. The higher up in the organization you go, the better.

Many companies actively discourage personal meetings, at least until a preliminary evaluation of the idea is made. The real reason for this is clear, although it will probably not be divulged to the person submitting the idea. To be frank, companies simply do not wish to waste the time of their high-priced executives on something which may prove to be old or impractical. The ploy most commonly used to resist a personal meeting at these early stages is to advise the submitter that it is impossible to have the right corporate people at the meeting without knowing the nature of the idea.

If you are interested in a personal meeting, and I think you should be, this argument can be effectively countered by letting the corporate contact know the general nature of your proposal. You need not get into specifics but simply provide just enough information about the idea to allow the corporate contact to identify the people in the company who should have an opportunity to consider it. A personal meeting to discuss your idea may provide you with just that edge required to earn acceptance by the corporate officials.

Don't forget: You and your idea are in competition with other people and their ideas. Selling an idea requires enthusiasm on the part of its creator, an enthusiasm that should be passed along to the people reviewing its merits. Don't leave the idea to struggle alone like an abandoned newborn on a doorstep. Get in there, if you can, to make your pitch. When you meet with people and talk to them, you learn their concerns and doubts. In a meeting you can, hopefully, counter them. If you're operating by mail, you never even get to hear objections which might be raised, much less have an opportunity to overcome them.

Whether you present your idea in person or mail it to the company, you will want to have a written presentation which can be distributed to the company officials. The presentation should reflect that you have given the matter a good deal of thought. It is often helpful to start out with a statement of the problem or problems the submission is designed to solve. The idea or invention should then be described with clarity, and the description should be of sufficient length to resolve any major unanswered questions that may arise concerning its nature or operation. Of course, don't overdo it. You don't want people nodding off.

If you've done market research, favorable results

should be included. Show your familiarity with the company's operations, and let the executives know how your concept will dovetail with these operations.

A few random, general thoughts about presenting your idea to a company: Once you have made the decision to make your submission, don't play it cagey. Be prepared to describe your idea completely. Don't hold back bits and pieces of information. A company must have sufficient information about your idea to permit decisions to be made. Being coy could kill your chances of making a deal.

Neatness counts. Your written presentation should look professional. If it's sloppy, the people you show it to will think you didn't care enough about your own idea to devote time and attention to it. Why should they? If your presentation is in a face-to-face meeting, look and act professional. Neatness counts here, too. Give the appearance of success and confidence. Look and act like you were on a job interview, like *you* were under consideration, not just your idea.

If you have a good, working model of your idea, by all means show it. Working models are invaluable sales tools. Be prepared to answer questions fully and honestly. Indicate your willingness to cooperate with the company in any further development work that may be necessary.

In the case of an unpatented invention or an idea, you will probably make your disclosure to the company after you have signed a release form which, as you've seen, usually contains a provision allowing the company to be sole judge of whether and how it will compensate you. Nonetheless, there is nothing to prevent you from indicating your wishes in the matter. A cardinal rule should be observed though. Exercise moderation. If your estimate of what your submission is worth is unreasonably high, you are likely to scare off a company that may have cooperated with you.

11
OTHER OPTIONS

OTHER OPTIONS

ONE OPTION—SUBMITTING YOUR IDEA TO THE UNITED STATES GOVERNMENT

You might want to consider submitting your idea to the United States government which, after all, has an operating budget that compares to that of even the largest company like an elephant compares to a flea. The federal government has a voracious appetite for new ideas and will consider all ideas and inventions submitted to it. Some departments actually scan patents at the time of issuance by the Patent Office in fields of particular interest to them. Each year the government receives many thousands of letters from individuals concerning ideas they feel might be useful to the myriad military or civilian agencies comprising our federal bureaucracy.

If you have an idea or invention that may be useful to the government, you may submit information concerning it directly to the military or civilian technical agency specifically concerned with products or processes of a similar type. In the event you do not know the proper agency that might be interested in your proposal, a situation not unlikely in view of the complexity of the federal establishment, you will have to do a bit of guesswork since, believe it or not, there is no central government agency to assist inventors in evaluating their developments. You might ask the office of your local congressman to help you identify the agency or agencies most likely to have an interest in your idea. Let your taxes work for you for a change.

Your disclosure to the government should certainly contain:
1. A statement of the advantages your idea possesses that make it superior to similar things in use or available
2. Complete information concerning the method or principle used in your idea
3. A description of your idea which is sufficiently clear to enable competent technical personnel to understand fully its construction and step-by-step operation
4. Any drawings, diagrams or photographs necessary to disclose the true nature of your idea
5. Any theoretical or actual performance data which you have showing the operability and superiority of your concept

In short, when dealing with a federal agency, you should be willing to make as full and complete a disclosure as possible to permit evaluation of your idea or invention, just as you would with a company.

The U.S. government discourages personal interviews with inventors, at least during the early stages of the evaluation process. Submitters are generally told not to come to Washington, D.C., for interviews, since the experts responsible for evaluating their ideas may actually be located in distant government laboratories. The arsenals, laboratories, and proving grounds where tests and evaluations are made are widely scattered throughout the United

States and occasionally overseas. The government feels that initial reviews should be handled by written correspondence.

Of course, you may not feel that way. I know I don't. As is the case with companies, I encourage personal meetings with federal officials, as high in the pecking order as possible, just as soon as possible, to promote a new idea. Good ideas slip through the cracks of government with far greater frequency than they do in the private sector—at least that's been my experience.

It is the policy of the federal government not to use any proposal in which an outside submitter has property rights without providing proper compensation. In addition, the government will, in the evaluation process, try to restrict information received from you to those persons having an official need for the information for purposes of evaluation. However, as a condition to receiving and evaluating your proposal, the government requires that it be clearly understood that acceptance does not imply a promise to pay, a recognition of novelty or originality, or a contractual relationship such as would render the government liable for any use of the information to which it would otherwise be fully entitled.

Of course, the government has many of its own personnel and those of outside contractors constantly working on research and development, and the substance of outside ideas and inventions may already be known to them or be in the public domain. For this reason, you will probably be asked to sign a disclosure agreement similar to those used by private companies.

SBIR Programs

The Office of Innovation, Research, and Technology, a unit of the U.S. Small Business Administration, administers programs aimed at helping small businesses meet federal research and development needs. The programs are called Small Business Innovation Research (SBIR) Programs and, under the right circumstances, these programs can be a source of funds to an innovator.

Each federal agency with an outside research and development budget in excess of $100 million must establish an SBIR Program. Presently participating agencies are:

Department of Agriculture
Department of Commerce
Department of Defense
Department of Education
Department of Energy
Department of Health and Human Services
Department of the Interior
Department of Transportation
Environmental Protection Agency
National Aeronautics and Space Administration
Nuclear Regulatory Commission

Each agency develops and publishes a list of research topics. Any for-profit small business concern with 500 or fewer employees can then submit a proposal for dealing with a particular topic. A proposal describes the results expected to be attained, the approach the firm plans to take, and how it will prove the feasibility of its approach. The agency then decides whether or not to accept the proposal, and, if it does accept the idea, it awards the company a sum of money (usually on the order of $20,000 to $50,000) to initiate research. Considerably more money (10 or more times as much) may be granted later if this initial research is found to have merit.

If you are interested in more information concerning the SBIR Program, request a copy of Publication No. SBIR T1 from the U.S. Small Business Administration, P.O. Box 15434, Fort Worth, TX 76119.

ENERGY-RELATED INVENTIONS

Another federal program exists for funding innovations, at least those of a certain type. The National Bureau of Standards has a department called the Office of Energy-Related Inventions, which will evaluate what it considers to be promising energy-related (non-nuclear only) inventions. If the evaluation is positive from the standpoint of either saving or producing energy, the U.S. Department of Energy might provide assistance. For more information, contact the Office of Energy-Related Inventions, National Bureau of Standards, Washington, D.C. 20234.

THE ENTREPRENEURIAL OPTION

Do you like a challenge? Are you ready, willing, and able to cope with a hornet's nest of problems? Do you enjoy hard work? Are you good at problem solving? Do you have business "smarts"? Are you innovative? Are you a risk taker?

If you can answer yes to all of these questions, you just might have what it takes to be an entrepreneur—

have what it takes to set up your own business for manufacturing and marketing your brainchild. Many people find this approach preferable to licensing or assigning their ideas. They enjoy having greater control over matters and the self-satisfaction that comes from shepherding a product all the way from inception to commercialization. They also enjoy the potentially greater financial rewards usually associated with entrepreneurship. Make no mistake, though. You will need drive and perseverance, plenty of each, if you tread the entrepreneurial path.

FINANCING YOUR VENTURE

In addition to the entrepreneurial skills and inclinations, you will need something else to get your venture going. That something is money. Check your pockets. How deep are they? Are they deep enough to get your new product and company off the ground? If they are, lucky you; but most of us aren't so fortunate. We have to look elsewhere for help.

Most start-up capital, the seed money needed to launch a new enterprise, comes from informal sources—from friends, family, ad hoc groups of acquaintances interested in inventing. At this stage of the game, unless you are someone with a proven track record, it's likely that you will only be wasting time and energy trying to get an infusion of capital from established conventional sources such as banks or venture capitalists. It's the old, old story of bankers being willing to loan umbrellas only when the sun is shining. You probably will have to obtain your seed capital from people who actually know you and have faith in you and your idea.

IDEA ANGELS

But there are other possible sources of start-up money. There are individuals who are "willing to take a flyer" on an idea if it is the *right* idea, even if the person with the idea is a complete stranger. In show business they're called angels. Maybe you can find your idea angel.

How do you go about this? Well, one way is simply to run a newspaper ad for an investor or partner. You see ads of this type in the *Wall Street Journal*, for example. Usually, because of expense, they are short, containing just enough information to attract an idea angel's interest. Your ad might simply state that you have a new product of a certain type or in a certain field and that you are looking for an investor. Be careful here, though. You can run afoul of Securities and Exchange Commission regulations if you are planning to raise more than a total of $500,000 with this approach. If you are thinking big money, talk to your lawyer before soliciting funds this way.

Another way of ferreting out potential investors is through investors groups and organizations. There are many associations of this type around the country and, if you're lucky, one might be in your specific area. These groups provide material support and assistance of all types for inventors and idea people generally. Obtaining seed capital is a common problem and you might find a member willing to give you specific recommendations about where to go and whom to see for financial support. Some of the more prominent innovators' support groups are:

Affiliated Inventors Foundation, Inc.
501 Iowa Avenue
Colorado Springs, CO 80909

American Society of Inventors
1710 Fidelity Building
123 South Broad Street
Philadelphia, PA 19109

Appalachian Inventors Group
P.O. Box 388
Oak Ridge, TN 37830

California Inventors Council
P.O. Box 2732
Castro Valley, CA 94546

Central Florida Inventors Club
4849 Victory Drive
Orlando, FL 32808

Confederacy of Mississippi Inventors
4759 Nailor Road
Vicksburg, MS 38180

Innovation Invention Network
132 Sterling Street
W. Boylston, MA 01503

Intermountain Society of Inventors & Designers
P.O. Box 1514
Salt Lake City, UT 84110

International Association of Professional Inventors, Inc.
Route 1, Box 1074
Shirley, IN 47384

Inventor Associates of Georgia, Inc.
637 Linwood Avenue, N.E.
Atlanta, GA 30306

Inventor's Association of New England
P.O. Box 3110
Cambridge, MA 02139

Inventor's Association of Washington, Inc.
P.O. Box 1725
Bellevue, WA 98009

Inventors Club of America, Inc.
Road 87
Andover, CT 06232

Inventors Council of Hawaii
P.O. Box 27844
Honolulu, HI 96827

The Inventors Council of Ohio
P.O. Box 37
Hilliard, OH 43026

Inventors of California
P.O. Box 158
Rheem Valley, CA 94570

Inventor's Workshop International
1781 Callens Road
Ventura, CA 93003

Midwest Inventors Group
Box 518
Chippewa Falls, WI 54729

Minnesota Inventors Congress
129 Bedford, S.E.
Minneapolis, MN 55414

Mississippi Society of Scientists and Inventors, Inc.
P.O. Box 100
Sandhill, MS 39161

National Inventors Cooperative Association
P.O. Box 6585
Denver, CO 80206

National Society of Inventors
23 Palisades Avenue
Piscataway, NY 08854

Northwest Inventors Association
723 East Highland Drive
Arlington, WA 98223

Oklahoma Inventors Congress
P.O. Box 53043
Oklahoma City, OK 73152

Society of American Inventors
P.O. Box 7284
Charlotte, NC 28217

Society of Minnesota Inventors
20231 Basalt Street, N.W.
Anoka, MN 55303

Tennessee Inventors Association
1116 Weisgarber
Knoxville, TN 37919

Texas Inventors Association
400 Rock Creek Drive
Dallas, TX 75204

Western Inventors Council
Oregon State University
Corvallis, OR 97331

Just spreading the word through business associates or social acquaintances can pay dividends. An inventor client of mine picked up a number of investors by telling his accountant about his need for start-up capital. The accountant did work for others with funds to invest and—voilà! Financial planners and money managers are other types of professionals who might be in a position to locate idea angels. If your banker is of a helpful nature and not overly conservative, he or she may be willing to steer you to an alternative source of funds.

INVENTORS EXPOSITIONS

Many states stage periodic expositions designed specifically to bring innovators and their new ideas into contact with companies and the public. If the right person or organization sees your idea at one of these events and likes it, you may have found your idea angel.

These expositions, which are frequently called by other names such as Inventors Fair, Inventors Congress, or Inventors Show, are often sponsored entirely by the state—usually the state economic or industrial development commission or a college or university. In some instances chambers of commerce, professional societies, or state manufacturers organizations serve as co-sponsors.

There is no standard program for an invention exposition. Each state develops a program which is best suited to existing needs and facilities. The exposition may be held alone or coupled with a products show, a procurement show, a technology utilization symposium, or other activity which increases interest and participation.

Practically all programs are set up, however, to

provide the inventor with an opportunity to do the following:

- Display inventions to potentially interested manufacturers, distributors, investors, and consumers
- Determine consumer reaction to inventions by demonstrating them to the viewing public
- Visit with other inventors and benefit from their experiences in handling their inventions

In some instances the expositions include symposia for participants covering such subjects as patenting, developing, manufacturing, and marketing new inventions and new products. They may also provide free consultation with experts in various fields pertinent to the protection, production, and promotion of new inventions. The expositions normally last two or three days.

Any inventor or patent owner may participate in expositions of this nature. In addition to private individuals, educational institutions, research foundations, and technically oriented companies as well as the creative employees of these organizations often take part. Prospective inventors or people who have ideas or inventions but lack knowledge about how to develop and promote them find the educational functions often associated with the expositions particularly helpful.

It is not normally a requirement of patent expositions that an inventor have a model of his invention. The display of a good, attractive working model is, however, often a valuable sales and promotional tool;

if you have such a model, you are urged to display it. In addition, photographs, film strips, and other information illustrating the operation of your idea and its advantages should be displayed. Also, it is often helpful to use as selling aids testimonial letters, awards, etc., pertaining to the invention. The ultimate sales aid is, of course, you yourself, and you should be prepared to answer questions about your idea.

The cost of participating in an exposition varies since a good deal depends on exactly how the exposition is financed. Generally speaking, the charge is nominal, covering the cost of the facilities being used and any banquets, etc., that might be a part of the official function. Established manufacturers exhibiting products, as is sometimes allowed, are usually required to pay a fee higher than that charged to private innovators.

Information concerning invention exhibitions that may be scheduled in your state can probably be obtained by contacting the inventors support group nearest you.

THE BUSINESS PLAN

Your Aunt Minnie may not give your business venture a steely-eyed look before investing, but an idea angel who doesn't know you certainly will. He probably will want to see a business plan before parting with cold, hard cash.

A business plan will set forth all relevent facts and projections concerning your planned company and the invention or idea it will be based on. This document should not be some slapdash effort with pie-in-the-sky projections. Not only do you not wish to mislead anyone, bear in mind that the plan will, in all likelihood, be carefully scrutinized by lawyers, accountants, and possibly others on behalf of the potential investor. If things don't make sense or add up, this will be readily apparent to someone sophisticated in these matters.

Business plans differ, of course, from fact situation to fact situation, but they conventionally include the following:

1. The company's name and list of principals involved
2. A summary of the plan
3. A summary of the product—what it does and its advantages—and its patent status
4. A marketing analysis, both past, if any, and projected sales
5. Production plans, including costs and capacities
6. Management—who they are and background
7. Financials—financial statement and plan, cash flow budget for a period of two years or so stating projected income and expenses, budgets, what the idea angel can expect to obtain from the deal

SBA ASSISTANCE

The U.S. Small Business Administration is a terrific source of services and information for the entrepreneur with a new idea. The agency has several programs that you might want to look into. Let's go through the most important.

Small Business Institutes (SBIs) have been established by the SBA at more than 450 colleges and universities around the country. The SBIs can help you by providing the market analysis and business planning assistance necessary to make your innovation successful. An SBI is the perfect place to go when you are preparing your business plan, for example.

How do you locate the closest SBI? It couldn't be easier. Simply look under the "U.S. Government" listing in your local telephone directory to find the closest SBA field office. That office will put you in touch with the most convenient SBI.

Courtesy of the SBA, you also have a chance to SCORE or even get in contact with a real ACE.

SCORE is the Service Corps of Retired Executives, a program staffed by retired businesspersons who volunteer their services. SCORE representatives will be happy to answer your business-related questions, either over the phone or at personal meetings—all for free. There are also SCORE pre-business workshops given on a regular basis for which there may be a nominal charge.

The ACE (Active Corps of Executives) program is similar except that the volunteers are active businesspersons.

For information about either ACE or SCORE, contact your local SBA office.

The SBA also makes available a large number of publications, very useful to the entrepreneur. They come in both pamphlet and booklet form and cover such topics as financial management and analysis, general management and planning, marketing and personnel management. For a list of these publications, write to the U.S. Small Business Administration, P.O. Box 30, Denver, CO 80201-0030, and ask for Forms 115A and 115B.

I mentioned earlier that the entrepreneur with little more than a new idea and a gleam in his eye is going to have a tough time getting start-up or seed money from a bank. With persistence, collateral, a good credit rating, and a firm idea of where you are going, you might be able to secure an SBA loan from a bank even in the early stages of your endeavor.

SBA loans are normally structured like this. The loan itself is made by a private lending entity such as a bank. The SBA provides a guarantee for the loan. Loans usually run for seven to ten years and are available at a variable rate. Because the SBA guarantees these loans, the lender's purse strings might be a little looser, but don't count on it if you don't have a good business-like proposal and a plan with a high probability of success.

12

The Dream Merchants

THE DREAM MERCHANTS

FRAUDULENT INVENTION PROMOTERS

A person with a new idea is like a proud new parent. And, just as it is difficult for new parents to be totally objective about their recent addition, it is often tough for a person who has developed a new idea or invention to be completely objective insofar as his intellectual progeny is concerned. More than likely, the kid's the greatest. The idea's the greatest. End of discussion.

There are exploiters around who trade on parental pride; to put it bluntly, they take advantage of it. Bogus talent agencies are one prominent example. In return for healthy sums of money they promise stardom in the movies or a career as a model. The woods are full of fraudulent invention promoters who operate the same way. For a fee these promoters promise all sorts of things. They will develop the idea. They will protect it. They will sell it. All it takes is money, some up front and some along the way.

Beware of these con artists. Avoid them like the plague. You will receive precious little for your money. The advice you get is likely to be bad advice. The services actually received can be obtained elsewhere—often cheaper or even free—and certainly with better results. You should know that invention promotion and marketing organizations are constantly under surveillance by both federal and state law enforcement agencies because of their fraudulent activities. Civil and criminal suits have been filed and continue to be filed against these outfits. A few states, such as California, Minnesota, and Virginia, have passed special laws dealing with some of the more nefarious practices, but even in those jurisdictions you have to be careful.

How Fraudulent Invention Promoters Operate

Before I tell you how the bad guys operate, let me again remind you that there are many reputable companies around to help an inventor evaluate, develop, and market his idea. I've discussed them earlier in this book and specifically identified some. In the discussion that follows, I'll let you know some of the things you should look for when you're shopping around for help so that you can separate the folks with the white hats from those wearing the black hats.

There appears to be a lot of money in "them thar invention hills," at least for invention promotion organizations. In recent years, there has been a virtual explosion in the number of individuals and companies offering to assist inventors to develop and exploit their inventions. It is pretty difficult to avoid encountering advertisements for these organizations, especially in newspapers and magazines likely to be read by people interested in science, electronics, or mechanics.

The ads, which often contain either an express or implied promise of surefire financial rewards to the inventor, offer such services as invention counseling, patent research, invention development and marketing, invention evaluations, patent searches and preparation, financing, franchise consultation, and so forth. These organizations are a nationwide phenomenon. Some are relatively small, even hole-in-the-wall outfits operating out of a post office box number, while others are virtual giants with branch offices located in major cities throughout the United States.

Regardless of their size or location, all fraudulent invention development and promotion companies operate in pretty much the same manner. An inventor responding to an ad will receive written correspondence and materials outlining the "services" being offered, often accompanied by a written agreement that he is asked to sign. Also, often enclosed is an impressive-looking form. The inventor is supposed to describe his invention on the form and return it. This packet of materials may be preceded or followed up by a phone call from a company employee, usually someone with a fancy title.

The proposed agreement normally provides that the organization will make an evaluation of the invention to determine its novelty and marketability. The inventor will be required to pay something for this, and the requested sum for "evaluating an idea" can run anywhere from $100 or so to, in at least one instance I am aware of, $2,000. In all probability, though, the initial amount that is requested will be relatively low. No use scaring the sheep away before you can fleece them.

The "evaluation" is referred to by some organizations as a market research investigation or feasibility study, but it all amounts to the same thing, a purported investigation of the merits of the idea. Some agreements might include a clause requiring the inventor to assign a percentage of any income that he might receive from the invention.

Let's say the inventor wishes to have the idea evaluated and returns the agreement, proof of invention form, and payment (of course) to the company. What happens after this? In all likelihood, after a relatively short period of time, a letter will arrive in the mail stating that the invention has "exceptional merit" or words to that effect. In some investigative work done by the California State Attorney General's office preparatory to bringing suit against certain invention promoters, investigators were unable to find even one instance of an inventor being advised that his idea stood little or no chance of being successfully commercially exploited. It's obvious that the process of evaluating inventions by these outfits is really nothing more than the first step in a fixed scenario dedicated to the extraction of a maximum amount of dollars from a new client.

Compare this practice to that followed by a reputable organization in the field. A company interested in really helping an inventor will make an honest initial appraisal at little or no charge. Because the appraisals are honest, the vast majority of inventions submitted to a reputable company are rejected.

An evaluation by a disreputable company is likely to have been made without even the benefit of a patentability search; or if a patentability search is made, it is generally a cursory one at best. Rather than concentrate on the actual merits of the specific invention which is the subject of the so-called study, the evaluation report presented to the inventor is more likely to incorporate superficially impressive statistics and other information leading the client to believe that a tremendous market exists for the invention which in fact may not be true at all.

For example, if the inventor's proposal relates to a reading lamp to be attached to a bed frame, the report is likely to contain statistics stating the number of beds used and sold in the United States as well as their total cost. Large figures of this type are usually bandied about in invention reports by fraudulent promoters to leave the impression that there is an absolutely tremendous market for the inventor's product once it is introduced into the marketplace.

In reality, of course, only a certain percentage of people employ bed lamps, and there are already many, many bed lamp designs on the market. The so-called evaluation studies will not mention any of these negatives. The types of information they incorporate in their reports, which are supposedly based on elaborate marketing studies, are in reality superficial compilations of facts from sources such as the U.S. Census Report, trade journals, etc. The reports are of very little value, designed only to leave the inventor with visions of royalty dollar signs dancing in his head. With the psychological climate created by these studies, it is little wonder that many would-be millionaires happily agree to the next step in the fleecing operation.

The evaluation report may suggest that the product is patentable. There may be an offer to prepare a patent application. The quoted charge will probably look cheap, at least compared to the estimates received from patent attorneys. There is a reason for this. An application prepared by an organization like this is apt to be slapdash. Not too surprising since there is more interest in getting your money than there is in obtaining a worthwhile patent for you. Another ploy is the offer, for a charge of course, to file a disclosure document with the Patent Office. Remember, this is something you can do yourself. No expertise needed there.

After the inventor is told that the invention is indeed meritorious and in fact looks like a real winner from a commercial standpoint, something else comes out of the bag of tricks. Upon payment of yet another fee to the company, it will assist the inventor in developing and marketing the idea. One piece of correspondence I have seen offered to provide these additional services for a period of one year upon receipt of a $1,200 payment from the inventor. The services could be extended for additional one-year periods by paying $600 per annum. These fees, of course, vary between the various invention promoters, and possibly within a single promoter's own operation. I cannot help but feel that the amounts of these fees are dictated to a large degree by how much the promoter feels can be extracted from a client's pocketbook.

What are these so-called developmental and marketing services? Generally, nothing more is involved than the preparation by the promoter of a brief description and drawing of the development. This material is then mass-mailed by the promoter to a number of corporations to see whether there is any interest in the idea. These presentations are often slickly designed and illustrated to give the client the impression that he is really getting something for the money.

Having worked for several corporations and having been on the receiving end of a vast quantity of these submissions, I can say with a high degree of assurance that shotgun mailings of this type by invention promoters are given little, if any, consideration by the companies receiving them. Most are not even given a reply. The promotional item ends up in the wastebasket.

A company is in fact much more likely to give consideration to ideas sent to it by individuals rather than by mass-mailing promoters. One reason for this is the fact that the larger companies receive unbelievable numbers of solicitations from invention promotion companies, most of which are directed to ideas or inventions of obviously dubious value. In fact, it is not at all unusual to find the same idea the subject of disclosures from a number of promoters on behalf of various clients, all of whom have been led to believe that they have come up with a hot new development. As you might well expect, ideas that have been thought of by this many people have also been given prior consideration by the experts within the companies that have received them. Because of the vast quantities of material of this type received by companies, even if an idea is worthwhile it will probably become lost in the shuffle with the reams of worthless material routinely submitted to corporations by invention promotion companies.

You would be well advised to contact a patent attorney before getting involved with invention promotion companies. The patent attorney is in a far better position to assist in evaluating the merits of an idea or invention and, in particular, to give advice concerning its possible patentability. Unscrupulous invention promotion companies often do not have anything in the way of expertise in this area despite their assertions to the contrary. Because of this lack of expertise and knowledge, there is always the strong possibility that an inventor will receive advice from invention promoters that is not legally sound. A failure to recognize problems and a lack of understanding of the patent laws and laws relating to other forms of intellectual property can seriously jeopardize an inventor's chances of getting the protection to which he is entitled.

DEALING WITH INVENTION PROMOTERS

Not every client takes my advice, and I suspect that there will be one or two readers of this book who will ignore my recommendation that invention promoters be avoided. And it is true that not all invention promotion companies are unethical or borderline con artists. If you develop an irresistible urge to deal with invention promoters, here are a few things to do to separate the good from the bad.

Before signing anything, or otherwise obligating yourself in any manner to a particular company, you should request a written statement from the company indicating in no uncertain terms the percentage of ideas that it has successfully marketed on behalf of clients. The promoter should also be willing to provide you with a list of these clients and a statement of the number of units of the invention that have been sold with the monetary return realized by the client from these sales.

You should also obtain in writing specific details

of what the company will do for you in return for the money you will give to it. In this connection, many of the agreement forms provided by invention promoters are fairly lengthy and couched in unclear terminology, such as a vague promise by the promoter "to undertake to utilize its best efforts to promote the invention." If you don't understand what is written, don't sign it.

One important thing to remember is that you should in no event rely on oral, as opposed to written, representations made by invention promotion companies. To avoid prosecution under state statutes pertaining to fraud, many invention promotion companies are reluctant to state in writing the many representations and promises they may make orally to you during a personal meeting. If an invention promotion company is not willing to back up what it says in writing, this should be a red flag. Terminate your negotiations with that company immediately.

Some companies will, for example, allege orally that they sell a large percentage of the ideas submitted to them for exploitation and that the inventors dealing with them receive a high return on their investment cost. Also, it is not uncommon to receive an oral representation that the company accepts only "money makers." You can guess how many of these companies are actually willing to back this up in writing.

Vanity Publishers

Inventors aren't the only ones who get taken. Authors have their own dream merchants waiting to take their money. The so-called "vanity publishers" are publishers that advertise for authors and promise to publish their books for a fee.

Authors are sometimes led to believe that they will end up getting more for their hard earned cash than they actually do get. For example, representations are sometimes made that certain marketing steps such as bookstore placement and advertising will take place. If these activities do occur, they are on a ridiculously small and ineffective scale.

In reality, what the author gets is a high-price printing job. You can publish your book a lot cheaper on your own. Just check around with commercial printers and compare.

SONGWRITING ASSISTANCE

Here's another area where the unsuspecting and unwary can get ripped off. You've probably seen the ads. Offers to help you with music and lyrics. Representations may be made that the company has an "in" with the recording industry, leading you to believe that the end product has a good chance for commercial stardom.

What you are actually likely to get is a slimmer wallet. Let's face it, if the people who prepare music for your lyrics or vice versa were all that good, they probably wouldn't be doing hack work on a mass production scale.

Sometimes the services advertised by these outfits include obtaining copyright registrations. As you now know, you can do this on your own at a fraction of the cost a songwriting assistance company will do it for.

Your chances of publishing a hit song or cutting a hit record through this approach are slim indeed.

IN CONCLUSION

Way back when, in Chapter 1 to be exact, I made a few promises to you.

I promised to provide plenty of useful information about how to develop, protect, and sell ideas. In a flight of rhetorical fancy I even went so far as to say that I'd make your journey through perhaps unfamiliar territory a painless, enjoyable one.

These many chapters later, we are at the end of the journey. In a way, though, your journey is just beginning. You, and your idea, are headed for a try at success.

It's up to you now. I've given you the tools. I've told you about the pitfalls to avoid. I expect, any day now, to hear that you've made it to the top.

APPENDIX

Let's assume that the paper clip has not been invented yet. You are the inventor and you have decided to file a patent application on this marvelous new device for holding paper sheets and similar objects together.

The drawings for your application for a utility patent might look like this:

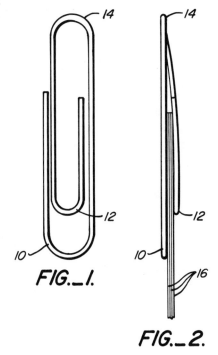

And the specification and claims something like this:

HOLDING DEVICE

FIELD OF THE INVENTION

The present invention relates to a device for holding a plurality of objects, such as sheets of paper, together.

BACKGROUND OF THE INVENTION

The problem of securing sheets of paper and similar objects together has been one of long standing and a number of devices have been developed for such purpose.

One prior art approach is the wire staple. The staple has several drawbacks. It mutilates the paper or other objects to be held together and requires a special piece of equipment called a staple gun. Then too, staples, once secured in place, are often difficult to remove.

Sheets of paper and similar objects have also been secured together by brads. Conventionally, brads are inserted through holes punched in the objects, a condition that is not always desirable.

SUMMARY OF THE INVENTION

Accordingly, it is an object of the present invention to provide a holding device for paper sheets and similar objects which does not require the mutilation of such objects.

It is yet another object of the present invention to provide a simple, inexpensive holding or fastening device which may be applied to paper sheets and similar objects without the use of auxiliary equipment and which may be readily removed by the user when fastening is no longer desired.

The device comprises a clip of unitary construction having one holder element positionable on one side of the secured objects and a second holder element positionable over the other side thereof. The holder elements are interconnected by biasing means which continuously urges the first and second holder elements together to lightly clamp the secured objects therebetween.

BRIEF DESCRIPTION OF THE DRAWINGS

Other objects and features of the present invention will become apparent from the following description taken in conjunction with the accompanying drawings in which:

Figure 1 is a plan view of a preferred form of the device constructed in accordance with the teachings of the present invention; and

Figure 2 is a side view of the device holding sheets of paper.

DESCRIPTION OF THE PREFERRED EMBODIMENT

Referring now to the drawings, a device constructed in accordance with the teachings of the present invention is illustrated. The device includes a first holder element 10 and a second holder element 12. The first and second holder elements are interconnected by biasing means 14 which continually urges the holder elements toward each other.

The device, as may be clearly seen in the drawings, is of unitary construction, being formed from a single elongate member, preferably metal wire. Other suitable materials, however, may also be employed, such as plastic materials.

First and second holder elements 10, 12 comprise looped sections formed at opposed ends of the wire and the biasing means 14 comprises a bight formed in the wire between the first and second holder elements. The looped sections are different sizes whereby one of the holder elements, in this case holder element 12, is adapted to nest within the other holder element under the urging of biasing means 14 when the device is not holding a plurality of objects together.

Figure 2 illustrates the positions assumed by the holder elements when the device is employed to hold a plurality of objects together. In Figure 2 the objects are paper sheets 16. It will, however, be appreciated that the device may be employed to secure other types of objects together as well. Objects are placed in holding position by pulling the holder elements apart and inserting the objects therebetween.

When the device is employed to hold objects, biasing means 14 is under torsion due to the inherent elasticity of the material of which it is constructed. First and second holder elements 10, 12 are continually urged toward one another, with first holder element 10 frictionally engaging one side of paper sheets 16 and second holder element 12 frictionally engaging the opposed, second side thereof. When the paper sheets are withdrawn from the device, holder elements 10, 12 will return to substantially the same plane with holder element 12 nesting within holder element 10.

CLAIMS

1. A device for holding a plurality of objects together, said device comprising:
a first holder element adapted to be disposed on a first side of said plurality of objects;
a second holder element adapted to be disposed on a second side of said plurality of objects in opposition to said first side; and
biasing means interconnecting said first and second holder elements, said biasing means continually urging said first and second holder elements toward each other whereby said first holder element frictionally engages said first side, said second holder element frictionally engages said second side, and said objects are clamped between said first and second holder elements.

2. The device according to claim 1 formed from an elongate member of unitary construction, said first and second holder elements comprising looped sections formed at opposed ends of said member and said biasing means comprising a bight formed in said member between said first and second holder elements.

3. The device according to claim 2 wherein said elongate member is a metal wire.

4. The device according to claim 2 wherein said looped sections are different sizes whereby one of said holder elements is adapted to nest within the other of said holder elements under the urging of said biasing means when said device is not holding a plurality of objects together.

ABSTRACT OF THE DISCLOSURE

A device for holding a plurality of objects together including a first holder element, a second holder element, and biasing means continually urging the holder elements toward each other.

INDEX

A
airline ticket ad space, 20
Apple Computer, Inc., 32–33
automobile windshield sunscreen, 33

B
"Baby on Board" signs, 33
bar code laser reader, 66
Barbie Doll, 19–20
Better Business Bureau, 46
"boomerang" letters, 73
Bureau of Commerce and Industry, 46
Bureau of Consumer Affairs, 46

C
Cabbage Patch dolls, 142
Carlson, Chester (Xerox Corporation), 29
Classification and Search Support Information System (CASSIS), 82
co-inventor, 102, 104
Coca-Cola bottle, 175
compact disc laser reader, 66
computers, anti-glare screens, 29–30
confidential relationship, 163–165
consumer survey, 57–59
contract
  express, 165–166
  implied, 166–167
copyright
  definition of, 93–94, 142–146
  fair use of, 155–156
  how to obtain, 147
  ineligibility for, 145–146
  infringement of, 17, 155, 156–157
  length of term of, 151
  licensing and assigning of, 157
  notice, 147–150
  registration of, 151–155
  semiconductor-chip protection, 157–159
  works of fiction, 17
Copyright Office, 94

D
Dahl, Gary (Pet Rock), 36–37
Definitions, U.S. Patent Classification, 78–81
developmental assistance, 46–48
Disclosure Document Program, 69–72
Disney characters, 142

E
Eastman Kodak Company, 170
entrepreneurial option, 206–217
  Active Corps of Executives (ACE), 216
  business plan, 214–215
  financing of, 207
  idea angels, 208

inventors expositions, 212–214
support groups, list of, 209–211
Small Business Institutes (SBIs), 215–217
Service Corps of Retired Executives (SCORE), 216

F
fads, 36–37
filling a need, 40–59
financial success, 12
"Find a Need and Fill It," 14
finding a need, 28–38
"first inventor," determination of, 62, 74

G
Garfield character, 142
Gould, Gordon (CD and bar code laser readers), 66

H
Hakuta, Ken (Wacky Wallwalker), 37
Haloid Corporation, 48; see also Xerox Corporation
Handler, Ruth (Barbie Doll, Nearly Me prostheses), 19–20
Homebrew Computer Club, 32–33
Hughes, Howard (rotary drill bit), 31
Hugster, 145

I
ideas
 aggravation factor, 34–35

commercial feasibility of, 52–59
consumer survey, 57–59
date of conception of, 62–63
describing, 63–68
developing, 28
developmental assistance, 46–48
employer rights to, 37–38
energy related, 206
fads, 36–37
financial success, 12
following trends, 34–36
forms of proof of ownership, 68–69
government and academic assisting organizations, list of, 50–52
making a model, 44–46
marketability, 55–57
myths about protection of, 72–74
novelty of, 40–43
private-sector assisting organizations, list of, 48–49
product evaluation, 53
proving ownership of, 62–74
public disclosure and patent protection, 59
reduction to practice, 74
rights to, 18
Index to U.S. Patent Classification, 78–81
purchase of, 81
"infinite-number-of-monkeys" paradigm, 144; see also Copyright

invention promoters, 220–228
inventions; *see also* Ideas
  one-year rule for patents of, 99
  patentability of, 96–98
  reduction to practice, 74
inventor's assistant, rights of, 47–48

J
Jefferson, Thomas, 92
Jobs, Steven (Apple Computer, Inc.), 32–33
Jones, Arthur (Nautilus concept), 35

K
Kitty Litter, 14–15

L
Lazer Tag, 36
Levi Strauss, 175
Lincoln, Abraham, 92–93, 113
Lowe, Edward (Kitty Litter), 14–15
LUV-BELT, 20

M
mail-order operations, checking competing products, 43
Manual to U.S. Patent Classification, 78–81
  purchase of, 81
Middle Market Directory (Dun & Bradstreet), 181
Million Dollar Directory (Dun & Bradstreet), 181
model making, 44–45

Murray, Gary (LUV-BELT), 20

N
Nearly Me prostheses, 20
needs
  filling, 40–59
  finding, 28–38
  where to look for, 31–33
notice of copyright, 148

O
Office of Innovation, Research, and Technology, 205
1-Two watch, 145
one-year rule, 99

P
patent
  applying for, 105–122
  assignment of, 128–131, 134–136
  attorney fees, 112–114
  cost of application for, 108, 112–114
  definition of, 93–94
  design, 122–123
  first issued, 92
  infringement of, 136–137
  joint ownership of, 131–132
  length of term of, 93
  licensing of, 132–136
  marking of, 138–139
  need for, 94–96
  one-year rule, 99
  patentability search, 41
  plant, 123–125
  requirements for, 98–101

sample application for,
  231–234
statute wording, 97–98
what can be patented,
  96–98
who may apply, 101–103
patentability search, 76–89
  at U.S. Patent and
    Trademark Office, 77–81
  outside Washington, D.C.,
    82–86
  reasons for, 76–77
patentability search services,
  86–89
  advertisements by, 88–89
  cost of, 87
patent agents, definition of,
  22–24
Patent and Trademark Office,
  19, 22, 41
  Classification System,
    78–81
  Disclosure Document
    Program, 69–72
  patent examiners, 80
  Public Service Center,
    address, 24
  Public Search Room, 77,
    79–80
patent application, cost of, 95
patent attorneys, definition
  of, 22–24
Patent Depository Libraries,
  82
  Classification and Search
    Support Information
    System (CASSIS), 82
  list of, 83–86
Patent Office Register of
  Patent Attorneys and
    Agents, 22
Peanuts characters, 142
Pet Rock, 36–37
pop-up thermometer, 35–36
prior art, 80
protection of ideas, overlap
  of, 172–175

R

Randolph, Edmund, 92
reduction to practice, of
  inventions, 74
*Reference Book of Corporate
  Managements* (Dun &
  Bradstreet), 189
Rubik's Cube, 31–32
Rubik, Erno (Rubik's Cube),
  31–32

S

Semiconductor Chip
  Protection Act, 158–159
Small Business
  Administration, 205
Small Business Innovation
  Research (SBIR) Program,
  205–206
  participating agencies, 205
songwriting assistance, 229
*Standard & Poor's Register of
  Corporations, Directors and
  Executives*, 180–181, 188
submitting an idea/invention
  to a company, 178–199
  patented, 185–189
  unpatented, 189–195
  how to disclose, 196–199

submitting an idea/invention to the U.S. government, 202–204

T
Tarkenton, Fran (airline ticket ad space), 20
Thomas Register, 43, 180
trade secret, 168–169
trademark
 definition of, 94, 169–172
 length of term of, 171
trends, 34–36
 health consciousness, 34–35
 throwaway concept, 35
Trivial Pursuit, 142
true inventor, 103

U
U.S. Patent and Trademark Office, *see* Patent and Trademark Office

V
vanity publishers, 228–229

W
Wacky Wallwalker, 37
Washington, George, 92
Watt, James (steam engine), 44
Williams, Ross (Wash 'n' Dri), 57–58
Wozniak, Steve (Apple Computer, Inc.), 32–33

X
Xerox Corporation, 29, 48, 179